THE POWER OF THE TALKING STICK

THE POWER OF THE TALKING STICK
INDIGENOUS POLITICS AND THE WORLD ECOLOGICAL CRISIS

SHARON J. RIDGEWAY AND PETER J. JACQUES

Paradigm Publishers
Boulder • London

All rights reserved. No part of this publication may be transmitted or reproduced in any media or form, including electronic, mechanical, photocopy, recording, or informational storage and retrieval systems, without the express written consent of the publisher.

Copyright © 2014 by Paradigm Publishers

Published in the United States by Paradigm Publishers, 5589 Arapahoe Avenue, Boulder, CO 80303 USA.

Paradigm Publishers is the trade name of Birkenkamp & Company, LLC, Dean Birkenkamp, President and Publisher.

Library of Congress Cataloging-in-Publication Data

Ridgeway, Sharon J.
 The power of the talking stick : indigenous politics and the world ecological crisis / Sharon J. Ridgeway and Peter J. Jacques.
   pages cm
 Includes bibliographical references and index.
 ISBN 978-1-61205-291-5 (pbk. : alk. paper)
 1. Indigenous peoples—Ecology. 2. Traditional ecological knowledge.
3. Indigenous peoples—Politics and government. I. Title.
 GF50.R54 2013
 363.7—dc23
                                                                                2013012879

Printed and bound in the United States of America on acid-free paper that meets the standards of the American National Standard for Permanence of Paper for Printed Library Materials.

Designed and Typeset by Straight Creek Bookmakers.

18 17 16 15 14   1 2 3 4 5

Contents

Preface: Lessons in Listening	*vii*
1 It Is Time for Our Hearts to Be Broken	1
2 Bretton Woods Takes Center Stage in the Carnival	23
3 *Industria* Encloses the Global Commons	47
4 Capitalism's Endless Pursuit of Profit Destroys the World's Food Supply	65
5 A Green Theory of the State	92
6 Earth Consciousness, Earth Action	118
7 A World Indigenous Movement: "We Are the Watchers, We Are Witnesses"	139
Notes	*155*
References	*179*
Index	*195*
About the Authors	*200*

Preface
Lessons in Listening

The Mi'kmaw people of contemporary Nova Scotia, Canada, have a tradition of using a "talking stick," and this tool is used to ensure that anyone who has something to say is given the space and deference to do so.

In *American Indian Quarterly*, Laura Donaldson describes it this way:

> Our Mi'kmaw ancestors used the Talking Stick to guarantee that everyone who wanted to speak would have a chance to be heard and that they would be allowed to take as long as they needed to say what was on their minds without fear of being interrupted with questions, criticisms, lectures or scoldings, or even of being presented with solutions to their problems. An ordinary stick of any kind or size is used.... The person who has a problem or issue holds the Talking Stick and relates everything pertaining to it, especially everything they have done to solve it. After they are through, they pass the stick to the person on their left, following the sun's direction. The next person, Nekm [the Mi'kmaw word for the next person], states everything they know about the problem without repeating anything that was already said.... The Talking Stick goes around until it returns to the person with the problem or issue, who then acknowledges everyone present and what they have said.[1]

Donaldson notes that the talking stick is an important facet of Mi'kmaw oral tradition, where deliberation and discussion, not the pen or the dictate, shaped social decisions. These discussions had to happen in smaller-scale groups than a nation-state or province, but it is a good starting point for thinking about fair and just deliberation as well as some of the lessons that Indigenous[2] peoples have learned over eons. The majority of this book offers a critique of the current world capitalist system and the way it is driving the planet beyond recognized boundaries, including its overall productive capacity and producing the Sixth Great Extinction through enormous land-use change, climate change, and serious pollution problems, to name a few challenges to sustainability of the human family.[3] The Sixth Great Extinction is a reference to the current extinction wave, which follows five other waves of extinction, such as the one that eliminated the dinosaurs, across the history of the biosphere.

We then ask that the power of this system be confronted, and that we allow for others to speak—and be heard. Indigenous peoples around the world have been mobilizing in a world social movement, calling attention to the global environmental threats to humanity, and to nonhumans. Few appear to be listening, perhaps because, if taken seriously, these voices could inspire broad social changes for the better that would be more cautious, less economically focused, and more sustainable. If the international community does not take Indigenous voices seriously, we risk throwing the human family into radical misery—joining many of our nonhuman families who already are suffering profound loss. The idea for this book was initiated by a dream one of the authors (Peter) had while traveling in the Intermountain West of the United States.

Peter was traveling with his two daughters, Olivia and Lily, taking them to see their family in Colorado, New Mexico, and Idaho. On the way back to Arizona, they stayed in a little family-run motel in southern Utah. That morning, he had a dream that interrupted his thinking about the human prospect. This dream was perhaps unlike any dream he had ever had—it was suffused with a profound sense of urgency and clarity.

> In the dream, I am watching people learn acrobatics and hang-gliding. Just like at a circus or a carnival, there was a giant safety net for the acrobats, who are bouncing very high. Some of the acrobats were permanently upside down, and they had painted faces or masks, some had jester-type hats on, and they all had colorful leotards of green, white, and reddish orange. There is a fence that people are standing in a circle around, on top of, and some are standing on the feet of the upside-down people. I think to myself that must be uncomfortable for the upside-down people. I want to try it bouncing on the net and doing acrobatics too, but I am not given the space and I eventually lose my turn.
>
> Later, I am at an American Indian jazz jam session in a large room. The jazz players were here in part because they had been excluded from the circus and had come here to do their own thing.
>
> I am given a coat from a medicine man to hold. His name is Scott. I took the coat without realizing that it could hurt me because it held powerful medicine. However, he makes it clear that he will not hurt me with it.
>
> The jazz session is over, and there is a heretofore unnoticed group of people who did not get to play. They have on the same leotards and costumes from the circus/carnival. They are demonic and mean to destroy us.

I hear Scott calling for his jacket, but I cannot find him. I accidently give it to another medicine man who is sleeping, but I take it back immediately. I knew the sleeping medicine man but cannot place him. I know we are in danger. I think I finally find Scott and return his jacket, but at this point I am a point-person for a quickly forming battle. The demonic carnival players were moving to destroy everyone from the jazz session and myself. We are now in a desert area with pinion and juniper trees, sage brush, and tumbleweeds. The earth—the soil—is red.

I face the demonic group, and I howl an unfamiliar sound that surprises me. I metamorphose into a white wolf. Then, I look behind me, and out from behind the sage and trees, tens of hundreds of wolves emerge, and we are one force. The sinister band also turns into wolves, and we fight them. They are preternaturally strong, but we know that this is not a simple battle but one upon which the future of life on Earth hangs. Together, we eventually defeat them. Someone says I had been bitten and my vision faded to black, but apparently I survived. I woke at this point.

I know this dream, like many dreams, sounds *more than* a little crazy. However, like many other dreams, I had my own sense of what the strange events meant. I knew immediately that I was being called to action to help defend the Earth against horrific forces. I believed that the carnival players were symbolic forces from the real world who wanted to distract us from the unfolding destruction of the Earth they were driving, and, as a person who was part of the group at the circus and who wanted to participate, I was complicit with, if not one of them. The American Indian jazz players and Scott were representations of wisdom and power of Indigenous peoples, though they had been cast out of the center of power, represented by the circus in my dream. Finally, I knew that the wolves were an authentic representation of fierce resistance, and that through my metamorphosis I discovered that we are not alone.

The central goal of this book is to call the other wolves—you—to join the struggle as a warrior with us. Another world is possible, and the vision exists to bring it to fruition, but these voices are being ignored while the powerful modern alliance of the state and concentrated capital are stripping down our rich and beautiful world, erasing the multitudes of nations in the web of life. It is our purpose to expose the current system for what it is and to suggest that ancient peoples of the Earth have something profound to say about alternative ways of being that hold promise for the human prospect.

Really listening well to others is one of our most important democratic deficits.[4] The essence of this book is a call to see what world powers are doing in our name and a call to hand the talking stick over to those voices who have been silenced by neoliberal powers. We are charged with the responsibility of listening and hearing what is being said, and that is the power of the talking stick.

Chapter One
It Is Time for Our Hearts to Be Broken

> The old people, surveying a landscape, had such a familiarity with the world that they could immediately see what was not in its place ... they went to work immediately to discover what this change meant.
> —*Vine Deloria Jr.*[1]

As we witness the accelerating devastation of our home, planet Earth, the time has come to acknowledge that the piecemeal approaches employed in the name of modern science to address this ecological destruction are failing. The authors of this book have spent years trying to support various technical solutions to stop further damage to the natural environment, yet conditions continue to worsen at alarming rates. In the United States, Congress has passed numerous forms of legislation attempting to protect our water, air, and nonhuman inhabitants just to see these laws weakened to the point of nonexistence as commercial interests declare them too costly. We are told that there is no alternative. We are told that we must drill for oil in ever-deeper waters even as oil spills, such as that of the 2010 blowout of the Deepwater Horizon in the Gulf of Mexico, break offshore oil spill records and cause untold damage, only to fade in our collective memory until the next one.

As ever more powerful and destructive hurricanes try to cool the oceans' rising water temperatures, we are told that economic growth demands that we burn even more fossil fuels, which spew greenhouse gases (GHGs), sulfur, nitrogen, mercury, and other pollutants into the atmosphere. GHGs absorb heat and raise the average global temperature of the Earth. This added heat has mostly been absorbed by the World Ocean. Between 1955 and 2010, the World Ocean soaked up 24 x 10^{22}J, or 24,000,000,000,000,000,000,000 joules of energy, and this added heat cannot be explained without including the effects of human greenhouse gas emissions.[2] To translate what this means, if this much energy were instantly transferred to the atmosphere, it would raise the average global temperature ~35°C (65°F).[3] Indeed, in recent decades, the Arctic Sea has lost nearly half of its ice.[4] In 2007, the Northwest Passage was open for the first time in human memory, allowing ships to pass through the Arctic. Tim Lenton has measured daily Arctic Sea extent since the availability of satellite measurements in 1979, and he comments, "The system has passed a tipping point."[5] Because the currents of the ocean are driven by temperature and salinity, the changing temperature is disrupting the way the ocean moves water around, which alters climate stability and disrupts the marine food web. That marine food web is, at the same time, being torn apart by overfishing and other pressures. We have depleted large fish (4 kg/8.8 lbs and larger) by 97 to 99 percent.[6] Overall, for every hook we put in the ocean today, we land half of the fish we did in the 1970s, and more and more fisheries are collapsing or are depleted, with almost none of the world fisheries recovering.[7]

There are signs everywhere we look that human use of the planet Earth is overwhelming her ability to survive in a form we can recognize. The great mammals of the sea are still being hunted "for science" while dolphins and other ocean mammals are beaching themselves for unknown reasons. In November 2012, the National Oceanic and Atmospheric Administration (NOAA) reported that October had finally ended sixteen straight months of above-average temperatures, making 2012 on track to be the warmest year on record since 1998.[8] The first ten months of 2012 were also the second most extreme on record in the United States, measured by assessing the top and bottom 10 percent in extremes of temperature, precipitation, and drought.[9] These broken records and extremes are consistent with global climate change as were the records set when post–tropical storm Sandy came ashore on the US coast on October 29, 2012, with the lowest air pressure reading observed in the Northeast, at 946 mb. Sandy flooded low-lying sections of New York and caused a record crest in the Delaware River, beating previous records set in 1950.[10]

Record global temperatures in 2010 during June and July also resulted in severe weather events, including massive forest fires in western Russia that lasted from July to September and unusually strong monsoons that caused deadly floods in Pakistan, all of which are now described with a very high level of confidence by scientists as being part of global warming.[11] Long-term warming of the waters of the Indian Ocean is causing floods in northern Australia while the southern-most regions are hotter and drier, causing drought. Fresh water is becoming dangerously limited due to warming temperatures that cause the ice packs of the world's mountains to melt so fast that great parts of the Earth will lose their water security. The eastern Himalayan Mountains, known as the water towers of Asia, have the largest glaciated area outside of the two poles. The melt from these glaciers supplies the river basins that provide fresh water for 1.3 billion people in India, Bangladesh, Bhutan, Nepal, and China.[12] Global climate change is causing increased rates of glacial retreat, threatening the water supply of these nations and causing drought and disruption of flow, which leads to climate refugees. Similar rapid mountain melt is threatening the water supply to the American Southwest from the Colorado River basin.[13] In the Andes, tropical glaciers are predicted to disappear within twenty years, threatening the water supply of 77 million people and roughly half of the hydroelectric power for Bolivia, Peru, and Ecuador.[14] The countries of the global South are set to feel some of the worst impacts of global climate change, yet they had virtually no role in causing it.

Philosopher of ecology Joanna Macy states it well: "We've been treating the Earth as if it were a supply house and a sewer. We've been grabbing, extracting resources from it for our cars and our hair dryers and our bombs and we've been pouring the waste into it until it's overflowing, but our earth is not a supply house and a sewer. It is our larger body. We breathe it. We taste it. We are it and it is time now that we venerate that incredible flowering of life that takes every aspect of our physicality." She continues, "World is lover, world is self and … it's ok for our hearts to be broken over the world."[15]

The most immediate threat to planet Earth is global climate change, yet the global North (wealthy, industrialized countries), especially the United States, has been unwilling to even make small steps to slow down the emission of GHGs. While China recently surpassed the United States as the largest emitter of GHGs, the global North is still responsible for the GHGs that have been building up since the Industrial Revolution. Yet, the countries that are hardest hit by the impacts are in the global South, including many Indigenous and subsistence communities who have produced minimal amounts to none of these gases.

Although there have been multiple global environmental conferences, experts indicate that "few conferences have yielded meaningful outputs that galvanized widespread state responses."[16] Perhaps most telling is the period of 1992–2012. One of the more important global conferences was the 1992 Earth Summit in Rio De Janeiro, Brazil, which announced the arrival of global environmentalism, launched several treaties, and heard dramatic intentions from world leaders. Twenty years later, the United Nations held the "Rio+20" conference, and it was clear that little, if any, genuine progress was made from the first Rio conference to the next. Worse, the urgent commitment apparent in 1992 was absent, with no important commitments or programs to emerge out of Rio in the summer of 2012.[17] The United States has the worst record of Annex 1 (wealthy industrialized) countries in that it failed to ratify the Kyoto protocol, which then expired in 2012 without any substantial treaty to take its place. The US Senate's failure to ratify Kyoto in the 1990s was then followed by the George W. Bush administration rejecting it totally. So far, the Barack Obama administration has done little to change this policy, seeking only voluntary limits of GHGs in secret negotiations at the 2009 Conference of Parties in Copenhagen and market mechanisms designed to benefit corporate polluters in Cancun, Mexico, in 2010, and virtual silence at the Rio+20.

Indigenous Voices of Resistance

Along the way, Indigenous peoples have not been silent, even if few have heard their coherent and consistent message. The peoples of the global South, especially the Indigenous peoples and subsistence farmers, who have been excluded from the mainstream international meetings led by nation-states, have still come together many times as heads of state alongside these official global conferences.

Indeed, across the last few decades, Indigenous demands have gone largely unacknowledged in mainstream media, domestic and international relations, and certainly in economic decision making. The latest of these Indigenous agreements was made in Rio de Janeiro in 2012 at the World Indigenous Peoples Conference on Territories, Rights, and Sustainable Development, or Kari-Oca II. The first Kari-Oca meeting occurred parallel to the state-led Rio Earth Summit in 1992, when Indigenous peoples from around the world met in their own conference and produced the Kari-Oca Declaration, discussed more in Chapter 7. "Kari-Oca" comes from the Tupí-Guaraní language, and it denotes the area where Rio de Janeiro now stands as "white man's house" and the settlements of the first Portuguese colonists. This second Kari-Oca meeting of Indigenous leaders from around the world was held *again* alongside, but outside, of the Rio+20 conference. Hortencia Hidalgo Cáceres (of the Aymara tribe), a member of the Indigenous Women's Network of Latin America and the Caribbean for Biodiversity (RMIB), captured the frustration of the Indigenous voices with the lack of action by nation-states. "We can't keep going on the same path that we have been on the last 20 years.... Real change is needed. We want to invite the world to a brighter future based on indigenous values and principles of 'buen vivir' (living well)."[18] Here, living well has nothing to do with economic growth, but is about "living in harmony with nature while pursuing material, social and spiritual well-being for all members of society, but not at the cost of other members or the environment."[19] The declarations produced by Indigenous leaders at Kari-Oca and Kari-Oca II have remarkable consistency with other declarations, such as the Cochabamba Declaration.

Indigenous leaders came together in Cochabamba, Brazil, in April 2010 because they were seeing the impacts of global climate change firsthand, and they wanted to warn the rest of us that there is no more time to wait. They came together to call for the creation of an International Climate Justice Tribunal, to press for a Universal Declaration of the Rights of Mother Earth, and to recognize the harm being done by the current capitalist system in which a handful of global transnational corporations are dominating all market exchanges. This World People's Conference on Climate Change demands that the countries of the global North, which have caused and continue to cause the buildup of GHGs, be held responsible to the rest of the world's people who will most suffer the consequences of global climate change. Not only have the developed countries not reduced their emissions since 1990, in fact, they increased emissions by 11.2 percent in the period from 1990 to 2007. The United States "has increased its greenhouse gas emissions by 16.8%, reaching an average of 20 to 23 tons of CO_2 per person. This represents 9 times more than that of the average inhabitant of the 'Third World,' and 20 times more than that of the average inhabitant of Sub-Saharan Africa."[20] Thankfully, the US per capita emissions declined in 2010—the data lag a few years—where US CO_2 emissions were around 18 tons per person.[21] The bottom line, though, is that rich countries have caused between 60 and 80 percent of global warming, while poor countries have contributed between 20 and 40 percent.[22] Further, most of these emissions from poor countries are from producing goods they export to wealthy countries.[23]

In a search for justice, Indigenous leaders have composed agreements that go beyond short-term technical fixes to not only global climate change but all impacts created by corporate-led economic globalization that are destructive to the natural world, which all forms of life depend upon. One particular focus is on the impacts of industrial agriculture, which is destroying their ability to feed themselves. The industrial approach to agriculture replaces diverse plants and animals with only a few chosen varieties that are grown for economies of scale while eliminating diverse Indigenous and peasant cultures along the way.[24] These forces threaten biological and cultural diversity directly and indirectly, "placing the world's diversity in both nature and culture increasingly at risk. *This means no less than placing at risk the very basis of life on Earth as we know it: the natural life-supporting systems that have evolved on the planet, and their cultural counterparts have dynamically coevolved with them since the appearance of Homo sapiens.*"[25] Areas of rich biodiversity and cultural diversity show "parallel extinction risk,"[26] where diversity is replaced with homogeneity "almost everywhere" and where "forces promoting homogeneity are playing an endgame on a global scale."[27] In response, Indigenous peoples and subsistence farmers seek food sovereignty in which there is a recognition of the right of peoples to control their own seeds, lands, food production in addition to the recognition of water as a fundamental human right.

At this point, we would like to pass the talking stick to the authors of the Cochabamba Declaration, an amazing document that captures many of the messages Indigenous leaders have tried repeatedly to convey to the rest of the world. In many tribal cultures, each member is given the right to address the rest of the village. The symbol of that right is referred to as the talking stick in particular North American tribes. The rest of the village accords the holder of the stick time to present his or her views. For the limited purposes of this book we present only a small portion of their statement, but strongly encourage all to go to the actual document to read it in its entirety. We now pass the talking stick to the Indigenous and other peoples largely of the global South, who begin,

> Today, our Mother Earth is wounded and the future of humanity is in danger.... The capitalist system has imposed on us a logic of competition, progress and limitless growth. This regime of production and consumption seeks profit without limits, separating human beings from nature and imposing a logic of domination upon nature, transforming everything into commodities: water, earth, the human genome, ancestral cultures, biodiversity, justice, ethics, the rights of peoples, and life itself.
>
> Under capitalism, Mother Earth is converted into a source of raw materials, and human beings into consumers and a means of production, into people that are seen as valuable only for what they own, and not for what they are.
>
> Capitalism requires a powerful military industry for its processes of accumulation and imposition of control over territories and natural resources, suppressing the resistance of the peoples. It is an imperialist system of colonization of the planet.
>
> Humanity confronts a great dilemma: to continue on the path of capitalism, depredation, and death, or to choose the path of harmony with nature and respect for life.
>
> It is imperative that we forge a new system that restores harmony with nature and among human beings. And in order for there to be balance with nature, there must first be equity among human beings. We propose to the peoples of the world the recovery, revalorization, and strengthening of the knowledge, wisdom, and ancestral practices of Indigenous Peoples, which are affirmed in the thought and practices of "Living Well," recognizing Mother Earth as a living being with which we have an indivisible, interdependent, complementary and spiritual relationship.[28]

The Hidden Costs of Economic Globalization

Before those of us immersed in a Western materialist consciousness will be able to truly hear these voices, we will need to unburden ourselves of the hubris of believing that we are the "civilized" peoples of the world and the rest of the world should try to become like us. We have not yet acquired a sense of humility that our ideas of progress have been constructed by a very specific knowledge system in service to a power mechanism that has now become globalized. Our lessons in humility can begin with the understanding that the Western form of "civilization" is destroying the planet Earth, even with only approximately 25 percent of the world's population engaging in this high level of consumption. We will begin our venture into a planetary consciousness by opening a space in this dense network of material consciousness through deconstructing the current world-system of corporate-led economic globalization.

Corporate-led economic globalization, the most recent cycle of collaboration between states and firms, is not what is generally thought of when one conjures up a free market economy of small firms described by Adam Smith as butchers and bakers competing for customers. Corporate-led globalization can best be understood as a state-sponsored global capitalist network created by selection of a few transnational corporations to hold quasi-monopolies in pursuit of endless profit. As a culmination of six centuries of global trade expansion to reduce the costs of production and ensure markets, states have used their power to privilege these firms in return for economies that produce economic growth. Industrialized states, mostly located in the global North, that are strong enough to pursue this strategy become even more powerful through their ability to tax the growing economy. This steady source of state revenues is then used to erect domestic and military bureaucratic mechanisms designed to maintain domination over the factors of production. The factors of production include labor, capital, and hidden costs of securing raw materials, disposal of toxicity, and infrastructure for transportation. The key to profit under this system is not efficiency, according to Immanuel Wallerstein,[29] but rests on the states' allowing these quasi-monopoly corporations to externalize these costs, especially the hidden costs. In simple terms, externalizing costs of production means that states use their bureaucratic mechanisms to help corporations pass a cost onto the taxpayer thereby increasing profit for the corporation.

For instance, under corporate-led globalization states help keep labor costs low for corporations by not imposing restrictions on runaway factories. These corporations are then free to go offshore and co-opt cheap labor worldwide. Additionally, strong states help secure capital for transnational corporations through international financial institutions such as the World Bank and the International Monetary Fund (IMF). What is generally not recognized by taxpayers worldwide is that they are picking up the tab for these costs plus the hidden costs of securing a constant supply of raw materials, the disposal of toxic wastes, and the building of infrastructure to transport the raw materials to the industrial core and products back to the periphery. This frantic race to rip raw materials from the earth, transport them to manufacturing centers, while leaving toxic wastes oozing from this rampage is now expanded to the farthest corners of the planet Earth. Having already depleted industrialized nations' raw materials at the core, this state-sponsored race for raw materials and cheap labor is now penetrating the most pristine ecosystems of planet Earth leaving a wake of destruction and decimating Indigenous and subsistence peoples who occupy them. This, at least in part, explains why diverse cultures and biological diversity face parallel extinction risks noted above, where linguists indicate that 60 to 90 percent of the world's 6,800 languages will be extinguished by the end of century, and around 4,000 of these languages belong to Indigenous peoples.[30]

How have the economic beneficiaries of this destruction hidden these impacts from us? They have convinced us that the products produced by this process are all essential to a modern concept of the "good life," and to the extent that any of this destruction becomes visible, we are told that it is a myth, an overexaggeration, or it is regrettable, but globalization is both the natural and inevitable course of progress. Consumption is the only god to be worshiped on this altar of unrestrained growth. The more stuff we have, we are told, the happier we will be. We are not even given the time to ask if a new product is needed, before we are prodded to race out and grab the latest version so we can be "better" than our neighbors. It becomes difficult to even conceive of who we might be if we resisted advertising screaming at us that we must have the latest iPhone, which is out of date by the time we walk out of the store.

What we wish to examine in this book is how this global economic structure is despoiling the planet Earth and decimating the most vulnerable human and nonhuman communities who must share the planet with these powerful political/economic actors. We wish to unmask the deceit perpetrated by the beneficiaries of this system and reveal the wizardry behind the curtain of this carnival of whirling lights and garish colors. We have been misled, and it is time in our planetary consciousness to wake up and decide for ourselves what the contours of a "good life" should enclose. We will not argue that there is a single concept of a "good life" that must be accepted by all. Far from this top-down version of "let's make a deal," we will present multiple "ways of being" in this experience of life on this planet. Corporate-led globalization is not the only vision and it certainly is not inevitable. The driving impulse of the authors is to facilitate the awareness that we all have the opportunity in this time to join with our neighbors, locally and globally, to define new concepts of a "good life." The normative or value parameters guiding these choices will be, first, an overarching concern to walk more gently on the planet Earth, always looking to take the least from her ability to maintain the web of life for the maximum amount of human and nonhuman entities now and in the future. The second parameter will be the recognition that humans are not only material beings but also spiritual beings who require meaning in our lives beyond how much stuff we own.

Traditionally, meaning has been guided by values embedded in our cultures that provide an understanding of who we are, what we are to do in this life, and how to understand our relationships with other entities. Today, the Western culture has been co-opted by the hegemonic ideology of economism, which embodies the capitalist imperative to pursue economic growth as the sole measure of value. To help us envision alternatives to a Western material conception of a "good life" sold to us by the carnival barkers of modernism, we will turn to the communities who have fought to resist this force. In particular today, we have the opportunity to hear the voices of Indigenous and subsistence communities who are fighting to maintain their own "ways of being" in the world. The voices of resistance that began this book are trying to warn us in the global North of the harm we are causing to the planet Earth, and are challenging us to simply allow them the dignity to pursue their own values and construct meaning for themselves. If we will but listen to their voices, they might help us unmask the carnival barkers and reveal the madness of this race toward global destruction.

Indigenous Ways of Being in the World

Many Indigenous peoples' long cultural memories include warnings to not repeat previous patterns of destruction. According to the Hopi, this is not the first time that the world has been on the edge of extinction. "The Fourth World, the present one, is the full expression of man's ruthless materialism and imperialistic will."[31] We—humanity as a whole—have emerged from three others after destroying them through malice and greed, discarding laws of nature that require respect for life. Malice, greed, and discarding laws of nature are the elements traditional Hopis see as the grounds for the end of this world as well, and as such they have social rules, an ethical system, and institutions that attempt to maintain their sense of the natural order and balance. Fittingly, the word Hopi actually means "peace."

Modern industrial institutions, such as the state and transnational corporations, view this kind of attitude as idealistic and even backward. This is why the mainstream expectations for relations between countries and corporations have been named "realist" or some variation, where self-maximization and aggrandizing power over others is a common goal. Yet, there is pathology in seeing exclusive egoism as immovable human nature and in justifying its patently unsustainable pursuits of expansive consumption and militarism against others. This pathology has established a very serious illness that requires healing, but at this point in industrial society there is even debate about just how real our environmental problems are, at the same time there is justification for wars of aggression (e.g., the United States in Iraq and Afghanistan), and the use of torture as a legitimate practice (again in the United States). This pathology has a kind of undeniable power to it. The United States, now the hegemon directing the world-system, is at the peak of its power militarily and economically. Even so, the use of this power has a cost and is showing signs of failing, and even recognized intellectual bulwarks of US power admit that the United States has a "systemic vulnerability to unregulated greed" and that the world arrangement of power no longer unilaterally favors the United States, even if this situation does not give the United States itself pause.[32]

Indigenous nations today, with their natural resources largely intact, are the primary targets of this kind of hegemonic power because they do not embrace self-maximization and aggrandizing power over others as a common goal, and therefore have not raided ecosystems in the same way modern powers have. Indigenous cultures are defined by an ancient and close connection to specific land that facilitates development of particular social, economic, and cultural customs and traditions in conformity with this connection to land.[33] Indigenous peoples are composed of existing populations who are descendants of peoples who lived on a specific land prior to the conquest or settlement of that land by other peoples.[34] Those peoples who have been able to survive conquest and genocide have been able to adapt to the ecology of their land, and have continued to maintain their own "way of being" in the world.

In 2000, the worldwide Indigenous population was 6 percent of the human population,[35] declining to 5 percent by 2009.[36] Between Indigenous groups, there is quite a variety in the extent to which each community has been able to resist the encroachment of the dominant state-sponsored economic systems and cultures.[37] Although many have integrated some elements of a market economy, even in its globalized form, into their communities, they have fought to maintain their way of life and the cultural values pivotal to it. Finally, we will be looking to those nations that are often termed the "Fourth World." These are "Nations forcefully incorporated into states which maintain a distinct political culture but are internationally unrecognized.... These are the 5,000 to 6,000 nations representing a third of the world's population whose descendants maintain a distinct political culture within the states which claim their territories."[38] These nations may or may not be Indigenous by the UN definition but what these nations have in common with Indigenous communities is colonial occupation and the desire for land they can call their own and over which they maintain sovereignty.

Critical to understanding these communities is the recognition that the very identity of the people is inseparable from their connection to land. Central to Indigenous identity is a communal relationship and responsibility to sacred spaces.[39] Unlike the liberal conception of private property that emphasizes the "right" to use your property as you choose, the communities of Indigenous peoples tend to emphasize the responsibilities to maintain the land in such a manner that benefits all current and future members of the community. Removal from cultural lands means the extinction of identity for Indigenous peoples. The vice president of the World Council of Indigenous Peoples, Hayden Burgess, put the stakes in no uncertain terms: "The Earth is the foundation for Indigenous peoples. It is the seat of spirituality, the fountain from which our cultures and language flourish.... Next to shooting Indigenous Peoples, the surest way to kill us is to separate us from our part of the Earth."[40] The corporate takeover of the world-economy is an attempt to do this by creating a "borderless world" in which there is a homogenization of land usage and cultural values thereby destroying any concept of community based on a connection, not only to a particular part of the Earth, but to the Earth herself.

Though not all Indigenous nations have been able to prevent the collapse of their societies, many have been able to maintain cohesive social groups through eons of time—thousands of years—in some very tough living conditions. In contrast, after only a few hundred years, and a very important last fifty years, industrial societies are threatened with collapse or wars of expansion to maintain their giant consumption at the same time these very societies alienate and dispossess Indigenous peoples from their land and their historic communities.

The book before you is not one of romantic idealization for Indigenous peoples, but one of respect for their experience that has the potential to guide the choices that face the modern world at this very moment. Indigenous cultures are not groups of people who were/are genetically predisposed to love the Earth and each other. This is the Rousseau-based myth of the "noble savage," the Indian who is always in harmony with nature. But, the "noble-savage" idea was, of course, a creation of the European mind (not just Rousseau's), where Indigenous peoples were still "savage" nonetheless, and a turn toward European-type "civilization" was both inevitable as part of human progress, and good. Indigenous people, therefore, who do not wish to move "ahead" and "develop," should, according to this modern paradigm, be brought into the modern world and its order for their own good. At the same time, this provides a basis for the dispossession of lands from Indigenous peoples for the sake of "progress." These lands are then available for exploitation by the institutions of capitalist globalization.[41]

The establishment of the modern institutions through the system of liberalism, and *neo*liberalism in particular, around the world is the essence of this corporate takeover, and some have argued that this face of globalization is the essence of human purpose.[42] Thus, in many ways, modern institutions and order are seen as "natural" and the inevitable destiny for humanity. Such an attitude is prevalent in today's global institutions like the World Trade Organization (WTO), the IMF, and the World Bank, but this is only the latest wave of globalizing Western ideals.[43] The previous waves started with European and Iberian global colonization efforts when the Western world expanded over the oceans and imagined everything they saw as somehow their own because they did not grant moral standing to Indigenous peoples, and therefore did not respect their tenure.[44]

This was the beginning of a world order that still exists today. This world-system is made of several expansionary and predatory elements that are designed to organize and funnel energy and wealth, which are the material bases for power, to the core or center from which this expansion started. It is within this context that we can explain the establishment of the state and the corporation, which are institutions designed—openly we might add—to concentrate wealth and power.

This sets up one of the most fundamental questions of our time—how does humanity sustain itself and the natural world that brings it meaning and possibility? Hopi prophecy warns that change is coming. For those who are divisive and care only for materialism, their world will end in similar fashion as the first three worlds—painfully, suddenly, and decisively. The Hopi prophecy also indicates that opportunists of the world will be left behind by those who have taken care of each other and the Earth:

> Those with peace in their hearts already are in the great shelter of life. There is no shelter for evil. Those who take no part in the making of world division by ideology are ready to resume life in another world, be they Black, White, Red, or Yellow race. They are all one, brothers. . . .
> The Emergence to the future Fifth World has begun. It is being made by the humble people of little nations, tribes and racial minorities. "You can read this in the earth itself. Plant forms from previous worlds are beginning to spring up as seeds. This could start a new study of botany if people were wise enough to read them. The same kinds of seeds are being planted in the sky as stars. The same kinds of seeds are being planted in our hearts. All these are the same, depending on how you look at them. That is what makes the Emergence to the next, Fifth World."[45]

Many sustainability thinkers understand that the world as it is cannot be sustained much longer, and that, like in the prophecy, change is coming. The hope is, however, that we can move to a new kind of world without staggering loss of life and planetary-wide suffering.

One option is to pay attention to what has happened in the past, and learn from this heritage to foster foresight.[46] This is what offers us some hope. Unlike the inhabitants of Easter Island, we have a sense of what caused their collapse, which resulted in desperate Malthusian misery in terms of starvation, violence, and much death.[47] We know now that Easter Island collapsed out of grim human-driven ecological destruction of their forests and biodiversity, which literally had fed them and allowed Easter Island to become a complex society. However, what is at stake today is the survival of a globally interconnected humanity, not just an isolated island. The choices modern world elites in transnational corporations, industrial states, and international organizations that serve this system make will direct the fate of billions of people—most of whom up to now have had little to no say in the direction of the world-system. While all corners of the planet Earth will be impacted by corporate-led globalization, the Indigenous and subsistence peoples, primarily living in the global South, are already experiencing the impacts now, and in the future will face even more severe water stress, soil erosion and decline, climate change, and massive species extinctions.[48] They will experience a continued enclosure of formerly public ecological spaces as they are fenced off for private accumulation of wealth. And they will bear witness to the demise of Earth others, or nonhuman nature, which continue to suffer as well, and this peril cannot be separated from humanity's own fate. They will have to face all of this without their traditional ways of being, including their traditional forms of agriculture that have evolved to withstand natural environmental stresses.

The global environmental changes we now are experiencing are beyond the Western historical record. Who, then, among us can we ask for lessons about how to choose between worlds? Indigenous people have this record. This is one reason why there is an active global resistance from Indigenous peoples to the liberal/neoliberal modern world—they have a record and are telling us that we are not on the right path. Hundreds of Indigenous groups have organized around the world with a similar message to change the dominant social order toward reverence for nature and a respect for people that disavows war and imperial expansion.[49] If we close our ears to their messages, we very well may lose the world we now have. There are, of course, no guarantees that there is an Arcadian Fifth World awaiting our emergence after we have consumed this one. Thus, we assume it is desirable and good to preserve our current home from global ecological decline.

Introduction to Planetary Consciousness

Once we have begun to recognize the harm being done by this takeover of the world-economy, the next stage is to deconstruct how our modern knowledge systems have blocked our ability to develop the humility necessary to hearing other planetary voices. A planetary consciousness begins with a new awareness of the connections amongst all life forms sharing existence on this planet Earth. In order to embark upon this path we need to understand how the dominant knowledge systems in Western modernity block forms of what might be termed "holistic knowingness." We will first examine how this knowledge system works to fragment awareness of our reality and then, secondly, we will turn back to the ancient knowledge system of philosophy to open a space in our imaginations to allow for the existence of a reality expansive enough to hear the voices of Indigenous peoples as a way to appreciate the gifts this planet Earth offers us and find our place within its welcoming embrace.

A preliminary step in unmasking the carnival barkers of corporate-led globalization is to name the enemy, which Amory Starr[50] contends is both a structural and discursive project. Structure refers to systems and patterns of social relations within which individual human choices (agency) are constrained. Discursive refers to the discussed, rhetorical construction of these structures and the framing of agency. For Starr, the enemy named is corporations because, she contends, their power has rendered states irrelevant. While we recognize that the power of corporations has greatly increased under globalization, we will follow Wallerstein[51] and William Hipwell[52] in identifying the state-corporate alliance as the enemy because, as will be discussed in Chapter 5, transnational corporations still need the state to secure ever larger markets and access to resources. These theorists do not agree on all aspects of this state-corporate alliance and we will point out when there are significant differences, but both present invaluable insights on how this alliance has been able to construct and maintain its power. Unless specifically referring to the work of Wallerstein, who has been developing the concept of a world-system since the 1970s, for the rest of this book we will adopt Hipwell's name for this alliance, *industria,* because it most clearly reflects the aspects of capitalism that are driving the global race for markets and resources that are at the heart of the panoply of harms done to the planet Earth and all who share her as home.

For Wallerstein, the structural component of this world-system is a capitalist world-economy with its roots in the sixteenth century. It is capitalist to the extent that the "system gives priority to the *endless* accumulation of capital."[53] With the aid of the state, firms set out to get more capital, simply to accumulate even more capital. Those who conform to this priority are rewarded by the system while those who do not are penalized by structural mechanisms and are eliminated from the social scene.

The unit of analysis in this world-system is the processes of production, not states. For Wallerstein, the world-economy gains its efficiency from "an axial division of labor between core-like production processes and peripheral productions processes which resulted in an unequal exchange favoring those involved in core-like production."[54] This means that there is a division of labor that channels wealth and power to the core-center (affluent, industrial states of the global North, which are often former imperial powers) from the peripheral states, typical of the global South. These peripheral states often had been colonies and today still face worsening conditions of trade based on raw materials, or now, some manufacturing. Because of these processes' dependence on the states, Wallerstein grants that it is not unreasonable to refer to core and/or peripheral states as long as one remembers that it is really the production processes that he is referring to.

Strong states have predominantly core-like productive processes formed by acting as patrons to quasi-monopolies, those who maintain their monopoly status for some thirty years. Because the system is not static, core-like productions often devolved into the periphery, such as the evolution of the textile industry, which could arguably be considered a preeminent production process in the 1800s and is now clearly a periphery process.[55] Weak states that contain mostly periphery production processes can do little to affect the axial division of labor in this capitalist world-economy.

Wallerstein's discursive component of this structure is a geoculture, which embodies the principles of liberalism beginning with the French Revolution and devolving into a much more volatile neoliberalism under the guidance of Ronald Reagan and Margaret Thatcher in the 1980s. The liberal program for the core countries of the modern capitalist world-economy was to establish themselves as "states based on the concept of citizenship, a range of guarantees against arbitrary authority, and a certain openness in public life."[56] The roots of this liberal geoculture are to be found in the French Revolution, which initiated the ideas that change was normal and sovereignty rested on the individual. Since that time, the geoculture for Wallerstein has been evolving in three contested areas—antisystemic movements, ideology, and the social sciences. Wallerstein maintains that most of the antisystemic movements have so far made only minor changes in the world-economy and have generally been co-opted as soon as they displayed any strength, such as the labor movements. In the arena of ideology, which concerned the pace and character of change in the social arena, the middle position of liberalism won over conservatism's fear of change from the old social hierarchical order and the radical's impatience with a lack of fundamental revolutionary change. This can be seen in the evolution of conservatism, where modern conservatism is a modification of classical liberalism, focusing on free enterprise and private property. The liberal ideology of change-as-normal soon became associated with the idea of progress and was adopted by the social sciences. The key institution to inculcate this ideology into future generations was the primary school that was to pursue science as the Rosetta Stone to creating a new civilization. "Science (replacing not only theology but philosophy as well) offered a path for material and technological progress, and hence for moral progress."[57] The divorce of philosophy from science and the subsequent reification of science as the epistemological standard led to a division in the study of social sciences. In the second half of the nineteenth century, the study of social reality became divided among history, economics, political science, sociology, anthropology, and "Oriental" studies. These disciplines, then, epistemologically had to move into either the humanities or natural science. Following the liberal ideology with its focus on the rational, the three disciplines that studied civilized social life including the market (economics), the state (political science), and civil society (sociology) eventually moved into a scientistic camp, while history, anthropology, and Oriental studies tended to consider themselves humanistic.[58] This division of labor among the social sciences is one of the primary problems that confronts us as we attempt critique the structure of *industria* and demonstrate its harms.

The deceptions perpetrated by the elites of *industria* only are able to succeed to the extent that they can keep us distracted. Illusions only are possible when the magicians can keep our attention focused on a small part of the movement, making sure we only see what they want us to see. This division of labor amongst the social science disciplines, which is supposed to help us understand our social reality, acts like a shell game that distracts our attention as we try to find the pea. A classic example of this is when neoclassical economics tells us that economic growth will cure all the problems of humanity and the environment. As we will show throughout this book, unrestrained economic growth is at the root of virtually all harms to the planet Earth. We will follow Wallerstein's lead and draw from all of the disciplines contesting their knowledge boundaries, because a planetary consciousness is not so much about presenting "facts" as it is a consciousness-raising exercise that requires more holistic thinking. The problems we have created are global in nature and fundamental to our very survival. We must begin to develop a consciousness that moves beyond intellectual boundaries, state territorial boundaries, and boundaries between human and nonhuman existence. We need a planetary consciousness that moves beyond ideological boundaries and focuses on how we are connected. Chapters 5, 6, and 7 are devoted to this endeavor.

We will freely draw on disciplines within both the scientific and humanistic camps and, even more importantly, from the lived experience of those on the front lines of the battle. We will use the insights from these sources to deconstruct the assumptions embodied in the geoculture of liberalism and, most particularly, neoliberalism with its uncritical promotion of unrestrained economic growth as the solution to all problems. While this will help expose some of the myths, it will not help us to explore *industria*'s illusion of what constitutes the "good life"; for this we need to turn back to philosophy and the recent work of Hipwell.

Several times we have introduced the concept that there are alternative "ways of being" in the world. This is more than adjusting some practices such as driving hybrids or recycling. It is derived from what philosophy calls "ontology." On an elementary level, ontology is what we believe the world to be, the essence of reality and existence. Is it like a machine in which atoms just randomly bang against each other or is it really more like an organism in which every element is aware of the whole? Under liberalism's divorce of philosophy from science, ontological questions are theoretically left to the individual, but are, in reality, embedded in unacknowledged assumptions of the liberal culture. The danger of defaulting on this level of consciousness is that we have become vulnerable to the shell game of *industria,* which is more than happy to place billions of dollars into advertising to tantalize and distract us into accepting a cultural homogenization based on Western materialism as the sole definition of the "good life."

Hipwell,[59] drawing on philosopher Gilles Deleuze and his collaborator, psychotherapist, and political activist Feliz Guattari, postulates a critical analysis of the geopolitics and conventional approaches to resource-use management[60] in *industria*. He defines *industria* as "a globalizing system of power/knowledge that has come to control most of the infrastructure of civilization."[61] This state-corporate alliance, in order to gain power and control over natural resources worldwide, has developed its own ontology and corresponding epistemology. *Industria's* ontology sees the "total field" of reality as discrete points that interact atomistically like balls on a billiard table. In this mechanistic ontology all phenomena must be placed into their own *identity* categories such as human races, academic disciplines, or plant and animal species. To fully create this "identitarian" ontology there is a mental or physical imposition of fixed, sedentary boundaries creating separate points out of a hitherto continuum of intensities that Deleuze and Guattari term "smooth space." Once everything has been separated, categorized, and given a name, the new "striated space" can be controlled by the state-corporate alliance. Today, we can see this very clearly in the "War on Terror." Once someone is labeled as terrorist, they become separate and no longer possess any status or rights that have to be recognized by the state apparatus. Similarly, once an element of a complex, wild ecosystem is labeled as a natural resource vital to *industria's* growth, it can be consumed without recognition of how it relates to its ecosystem or the larger planet Earth. For instance, the George W. Bush administration in the United States greatly accelerated a form of coal extraction called "mountain top removal" in the Appalachian Mountains in which coal companies blast off the tops of mountains, allowing the chemicals and wastes to fall into the rivers at the foot of the mountain. The mountains are no longer seen even for their utilitarian value as watersheds or aesthetic inspiration. They are reduced to merely unimportant containers of the resource coal, and the people who depend on the streams for their water supply now only find black water flowing from their taps. Clean water does not yet fit neatly into a category so under "identitarian" ontology it is simply marginalized and dismissed until it becomes useful to *industria*. This marginalization of water, vital to human survival, is also marginalized in the practice of hydraulic fracturing, or "fracking." Governments along with gas companies are pushing headlong into the controversial process of releasing natural gas from underground reservoirs by fracturing the surrounding geology with millions of gallons of water along with toxic chemicals, and when a majority of the water and chemicals returns, it often comes with radioactive contaminants as well.[62] Here, the probabilities for contamination of expansive water supplies are being recklessly threatened for profit literally by the mechanical tentacles of globalist desire, while waving away the concerns of citizens who want to protect a vital element of life. Once fresh water is scarce enough to generate enough profit, *industria* will no longer marginalize it.

All ontologies privilege an epistemology (a way of knowing about reality), and "identitarian" ontology is no exception. The creation of fixed, sedentary boundaries facilitates an epistemology based solely on instrumental rationality, which privileges a logic that can aggregate, measure, and calculate all social problems. Poverty becomes measured in income per capita, as does happiness; if only you can make a certain amount of money, you will be happy. The problem of course is that this amount is always a moving target as it is a relational measure to the amount of money others are able to accumulate. There is no sense of enough. This limited, instrumental rationality is very useful to the separation of the academic disciplines, each deriving their "facts" to demonstrate that their measures of the "good life" are more accurate for the economic human, political human, or social animal. Deleuze and Guattari maintain that this instrumental rationality is at the core of what has been globalized. They term this "state-thought," which is much more than the state system and encompasses "the entire suite of administrative, disciplinary and technological strategies that gave birth to states in the first place."[63] Moreover, it is "organized on the model of the tree: vertical, hierarchical and comprised of isolated branches of knowledge between which communication and exchange are difficult."[64] Hipwell concludes that *industria*, initially gestated in the state system, and now, "armed with State-thought, and backed by State military power(s) and territorial claims . . . has been born to take its place as ruler of the world."[65]

In contrast to "identitarian" ontology, Deleuze contends that the real world is not composed of static identities but is always fluid and mobile. Nor is reality homogeneous. It is ontologically characterized by "difference." Hipwell explains, "This difference is not as it might seem, difference *between things* (an identitarian notion) but rather the idea that reality is a continuum of interplay, interpenetration and interconnectedness and that 'things' are merely *intensities* in this continuum, internally constituted by the interplay of different forces, and themselves interacting and interpenetrating with everything around them."[66] In this ontology of "smooth space," social reality and the world are viewed holistically. All components of the web of life are acknowledged as an integral member of the larger web upon which they depend. For anyone who has played or watched team sports it is not difficult to understand the role of individual players in relationship to the team as a whole. Whereas every team member's contribution is recognized, for a team to win, no individual member can replace the importance of the functioning of the team as a whole. In an ontology of smooth space, each energy has a unique identity but it is also an integral part of the larger whole upon which it is dependent for life. The identitarian ontology of difference between things obscures humans' interdependence of all elements of the biosphere, as if humans could live without air, water, or the smile of a friend.

The epistemologies of instrumental rationality and logic, while useful, are unable to grasp these connections among difference inherent in the holistic nature of reality. Deleuze maintains that he is an empiricist and materialist and therefore does not dismiss the usefulness of logic and reason. However, he holds that these epistemologies cannot grasp the holistic nature of reality. To help us grasp this holism he urges us to reconsider a much misunderstood epistemology that he calls wisdom of the body, in which he calls for a rehabilitation of intuition. Hipwell states, "Prior to intellectualization, our instincts respond to and process the sensory data received by the body as a continuous and multiplicitous flow."[67] Often, this intuition can come in the form of a dream or meditation as in the genesis of this book. When we pass the talking stick to Indigenous voices in Chapters 6 and 7, the reader will be able to "hear" alternative ontologies and their corresponding epistemologies from those who live them. It will be clear as you listen to their stories of reality that they are not based in an identitarian ontology of separation, which would deny their deep sense of connection to all life.

So, how can Hipwell's concept of industria help us unmask the swindlers of corporate-led globalization? On the most fundamental level it will allow us to address the unacknowledged ontological assumptions implanted into our subconscious by the modern discourse of liberalism and, particularly, neoliberalism. Neoliberalism strips away all noneconomic forms of relationships, leaving us with only market mechanisms to account for interactions amongst peoples of the world, humans and nonhumans, and the entire planet Earth. Relationships as vastly varied and unique as parents' love for their child, communal ties within an Indigenous tribe, or a top predator's vital role in an ecosystem simply do not exist under a neoliberal model of reality. Under neoliberalism, rich cultural practices are stripped of their meaning, citizens of states are reduced to consumers maximizing their utility, while the rich, dense complexity of ecosystems is erased through dissection of the whole into parts valued only when they are ripped from their ecosystem and commodified for exchange in the market place. The sole relationships that are acknowledged are those that can be measured as economic exchanges.

Neoliberalism's market model of all relationships comports perfectly with identitarian ontology in which all elements are separated, categorized, and given a name. Each element, whether a human or a tree, is placed into its own identity category. With all of reality so nicely separated, the only interaction posited is atomistic, mechanical interaction of parts, like balls on a billiard table. There is no whole, and component parts are only valued materially in that they are economically useful to the globalizing power/knowledge system Hipwell calls *industria*. The epistemology privileged under identitarian ontology is equally as abstract and reductionist. Under this ontology it seems perfectly rational to equate a computer model of an ecosystem with the immeasurably complex relationships actually operating in the ecosystem. This reductionist epistemology would be challenged in an ontology that acknowledges a rich complexity of relationships amongst unique elements of reality. The assumptions undergirding the ontology of "smooth space" militate against reductionism by acknowledging the continuum of interplay and interconnectedness of unique, but not separate, entities of reality. "Life forms are not discrete but are all to some degree interdependent and interconnected."[68] Epistemological reductionism in which all life forms of reality become abstracted from their context can only be theorized in a mechanistic ontology. In a mechanistic ontology, relationships are only recognized if they are based on physical force. Love or other human sentiments are considered merely subjective or abstractions not worth considering.

Locked into the assumptions of identitarian ontology and its privileging of Western logic and instrumental rationality is the denigration of alternative ontologies and epistemologies as primitive. Yet, it has been the modern scientific epistemologies that have facilitated the destruction of planet Earth's life support systems that we see emerging today. This logic has produced forms of chlorinated molecules (chlorofluorocarbons or CFCs) that are transported to the stratospheric ozone layer where they act as catalysts to break down ozone (O_3) to oxygen (O_2) thereby destroying the shield that protects the earth from cancer-causing UV rays.[69] The economic usefulness of these CFCs in everything from hairspray to Styrofoam cups and refrigerators made factories pump them into the atmosphere at accelerating rates without any caution about how they might interact with the biosphere's fragile atmosphere. Under modern scientific logic that only looked at individual, separate applications of the new and miraculous nontoxic, nonflammable coolants,[70] it was perfectly rational to produce and use them every way possible in order to promote a modern scientific idea of "progress," which since Francis Bacon has been understood to be anything that facilitates human control of nature. The mechanistic ontology that privileged this epistemology did not posit reality as a continuum of interplay, interdependence, and interconnectedness that would have cautioned against rapid introduction of such a powerful new force into an existing, precarious equilibrium. What has turned out to be "primitive" is the hubris of modern scientific knowledge, which is based on identitarian ontology of separateness, denying the recognition of holism.

Hipwell's and Wallerstein's insights on corporate-led globalization will additionally guide our analysis by reminding us that *industria* has both a physical structure and an accompanying discourse that justifies its power. In the following chapters we will deconstruct both of these components. Chapter 2 begins our deconstruction of the role of the international financial institutions in the construction of *industria*. We begin with the first two Bretton Woods (a resort in

New Hampshire) institutions, the World Bank, and the IMF, which emerged as the backbone of this state-corporate alliance. These international financial institutions were erected by the core states, especially the United States and Britain, after World War II to help corporations replace the mechanisms for control of global natural resources and markets that had existed under colonialism. The discourse that arose to justify this takeover of noncore states' resources was called "development discourse." In the name of modernization and progress, it was necessary for first the World Bank and then the IMF to help these poor, "underdeveloped" states. In reality, the so-called development allowed Western corporations virtually unhindered access to the natural resources of these countries, while creating a dependency for industrialized goods manufactured in core states.

Chapter 3 adds on to our understanding of how the neoliberal agenda emerged in the early 1980s and was further institutionalized worldwide. While the World Bank and the International Monetary Fund had been able to strip the ability of Third World nations to protect their newly emerging industrial economies, core states, especially the United States under Reagan and England under Thatcher, needed a new stronger entity to complete the institutionalization of a global neoliberal agenda, in which the only model of development accepted would be economic growth: the "rising tide lifts all boats" model of how to reach the good life for all planetary citizens. For this purpose, the General Agreement on Tariffs and Trade (GATT), the third Bretton Woods institution, was morphed into the WTO. Any policy that is not in accordance with the neoliberal market model can be ruled as a "barrier to trade" and, therefore, WTO-illegal.

Chapter 4 takes a deeper look into the impact of these neoliberal policies on the ability of Indigenous and subsistence peoples to feed themselves. Industrial agriculture, pushed by transnational corporations, is literally destroying the ability of farmers to decide what food they grow and how to grow it. They are being forced to use genetically modified organisms (GMOs), seeds that have been created in a laboratory, in place of native seeds they have been evolving over hundreds of years to suit the climate and traditional diets of the people and that don't require expensive chemicals or water to grow. Many of the GMO seeds are wind pollinated and will forever destroy their own unique seeds. However, if they do not buy these GMO seeds and all of the chemicals, they are prevented from selling their products outside of their immediate area. Moreover, consumers worldwide do not have a full range of healthy and safe foods available in the local supermarkets or, better yet, buy from local farmers' markets. Instead, we get the apples that won't bruise and look nice on the shelf, but have little taste and very little nutrition. To get some flavor into supermarket processed foods, corporations load them with fructose and salt.

In Chapter 5, we make explicit the role of the state in this state-corporate alliance and put forward a suggestion of how citizens, especially in core states, can join in a planetary consciousness with those in the Third World to resist this corporate takeover of our "ways of being" in the world. We put forward a green theory of the state in which it is explicitly acknowledged that state power is directly dependent on the use of natural resources.

In Chapter 6, we pass the talking stick to allow Indigenous voices to speak for themselves, in a language of their own. Perhaps if we can begin to listen, we will recognize that there are alternatives to corporate takeover of the global economy. Chapter 7 introduces some principles basic to an emerging planetary consciousness based in respect for all life on planet Earth that is fundamental to all Indigenous traditions.

People are not simply utility maximizers but rather live in families, communities, cultures, and very different places on planet Earth that are recognized, at some level, as home. If we do not reclaim the meaning inherent in these unique and rich differences, we humans, along with the rest of reality, will be converted into an identitarian category to be done with as *industria* decides. We have to reclaim the right to define our own way of being—first by recognizing that we have one.

Chapter Two
Bretton Woods Takes Center Stage in the Carnival

> The Ogoni and other peoples of Nigeria are waging nonviolent resistance against Shell, one of the most powerful multinational corporations in the world, as well as against the International Monetary Fund and the oil-rich, corrupt Nigerian government—to reclaim our lands, our livelihoods, and our homes.
> —*Oronto Douglas and Ike Okonta (Ijaw)*[1]

The structure of *industria*'s modern form began to emerge out of the ashes of World War II. With the threads of traditional colonialism beginning to tear, new mechanisms had to be developed to assure the victorious states continued access to the world's natural resources and markets for their finished products. The primary bureaucratic apparatuses set up by the victors are collectively referred to as the Bretton Woods institutions, which include the World Bank, the International Monetary Fund (IMF), and the General Agreement on Tariffs and Trade (GATT). GATT was transformed into the World Trade Organization (WTO) in 1995.

A primary concern after World War II was the stability of the international financial system. To ensure this stability, the IMF was set up "to monitor currency values and help avert balance-of-trade crises, and the World Bank was meant to make loans to re-develop war-torn countries."[2] During the world depression of the 1930s, trade wars had broken out as countries defaulted on loans. From 1914 to 1930 more than $17 billion worth of foreign bonds were floated in the United States, in addition to bonds by almost every enterprise imaginable. "By

1935, 38 percent of the foreign bonds held by Americans were in default."[3] As American investors began to withdraw from foreign markets, countries tried to undercut each other by devaluing their currency and erecting trade barriers. These new Bretton Woods institutions were sold as necessary to prevent a reoccurrence of these ruinous trade battles, arguing that trade and civilization were inseparable, making these institutions essential to global prosperity.[4]

Today, it is clear that the original organizing principles of a stable global economy and prosperity for all have been diverted into profit for transnational corporations (TNCs) and a global elite that benefit from them, to the exclusion of a system based on shared justice and environmental sustainability.[5] As will be demonstrated throughout this chapter, these Bretton Woods institutions have morphed into mechanisms that have permitted a small minority of developed states, working today on behalf of TNCs, to reach into and control the internal economic and social policies of countries in the developing world.[6] This new system has become central to what is often referred to as the New World Order, a term first enunciated by George H. W. Bush, drawing on the work of a Washington-based right-wing think tank, the Heritage Foundation. The New World Order, founded on neoliberal principles, contends that unrestricted economic growth is the sole path to prosperity. If all state barriers are removed, it is argued, TNCs will create a global trade regime that will insure economic growth. Moreover, once there is a sufficient amount of economic growth there finally will be enough prosperity to address questions of social justice and protection of the planet Earth. However, no amount of "sufficient" in this neoliberal heaven is defined and, in the meantime, all of the benefits go to a new international capitalist class.[7] After twenty years of neoliberalism on a global scale, the flaw in this New World Order was clearly articulated by a United Nations Development Program report, which concluded, "Economic growth cannot be accelerated enough to overcome the handicap of too much income directed to the rich. Income does not trickle down; it only circulates among elite groups."[8] The TNCs, and the elites that control corporate-led globalization in both the developed and Third World, skim off the profit of economic growth while the majority of the world's population remains in stagnant or declining economies and deteriorating social and environmental conditions.

This New World Order accurately reflects Hipwell's characterization of a state-corporate network designed to increase its own power at the expense of planet Earth and all those not directly aligned with the network. This chapter will examine how *industria* emerged as a structure, what justification was used to obscure its true intent, and, finally, explicate the impacts of its avarice on the developing world, especially on the poorest in the world and their natural environment. In this chapter, the majority of the focus will be on the World Bank and the IMF. Chapter 3 will examine trade under GATT, its transformation into the World Trade Organization (WTO), and, in Chapter 4, we will focus on subsistence agriculture as practiced by many peasant and Indigenous peoples. Understanding how power is exercised through these institutions is a vital step in confronting the domination of *industria* and its corporate sponsors. A power cannot be resisted until it is visible. Let us begin with who the players are and how they pull the strings.

American and British officials met in secret in the fall of 1943 to conduct meetings on how to stabilize differing national currencies and insure prosperous trading partners after the War. Europe was lying in rubble, and the countries of Asia, Africa, and Latin America were not up to world market standards. When news of the meetings leaked out, the sessions were ended, but were resumed by President Franklin Roosevelt in 1944 at the Mount Washington Hotel, near the resort of Bretton Woods in New Hampshire. More than four hundred delegates from forty-four Allied countries, led by the United States and Great Britain, drafted a charter and rules for two international financial institutions, the International Bank for Reconstruction and Development (IBRD, known more informally as the World Bank or simply the Bank), and the International Monetary Fund (IMF). From their inception these two state-based institutions were linked. No country could participate in the World Bank's lending programs without first joining the Fund.[9]

The World Bank and the United States

While the Bretton Woods institutions maintain a formal stance of independence from political or ideological influences of member states, it was clear from the beginning that the Western developed economies, especially the United States, would play a decisive role. In the case of the Bank, the "United States is the Bank's most powerful member."[10] Initial mechanisms of control were instituted by situating the Bank's headquarters in the United States and establishing English as the official language. The United States is also the largest single shareholder and is still the only country with veto power over amendments to the Bank's the Articles of Agreement. More importantly, by tradition the leader of the Bank has always been selected by the United States and has never been reversed. Even the extremely controversial George W. Bush appointment of his former Deputy Defense Secretary Paul Wolfowitz was not blocked, although he was widely opposed by those professionally engaged in development, including the staff of the Bank itself.[11] The Obama administration continued this dominance with the somewhat controversial nomination of US citizen Dr. Jim Yong Kim in April 2012. The G20 and heads of governments had committed to a more transparent, merit-based selection process, but with the support of the European members of the Board of Directors, the United States was able to push through their candidate. Another highly qualified candidate, Ngozi Okonojo-Iweala, a former finance minister of Nigeria and managing director of the Bank, emerged as the preferred candidate of the developing countries and would have allowed president Obama to be the first to appoint a woman from a developing country, but with the help of the Europeans, who had received support from the United States to maintain control of the IMF when Strauss Khan was forced to resign earlier, the global North maintained their control of both international financial institutions.[12] Each president of the Bank has an official position of neutrality, but as a Reagan Treasury Department report, which was critical of the Bank, reluctantly concluded, "the Bank usually acts in support of the long-term American political and strategic interests."[13]

A second source of US control comes from its vigilant oversight of World Bank loans and policies, which, of course, is facilitated by the Bank's location in Washington, DC. "Alone among the Bank's members, the U.S. reviews each loan proposal in detail, and officials of the Treasury Department are in daily contact not only with the U.S. executive director but directly with Bank officials."[14] Congressional influence from this close scrutiny shows most when the Bank needs to replenish monies for its programs. The Bank has three primary programs: the original IBRD, and two later programs, the International Development Association (IDA) for more impoverished countries, and the International Finance Corporation (IFC), which was set up to provide loans to the private sector. The IBRD originally limited its financing to countries for specific, clearly defined development projects, which have historically been interpreted by the Bank to be large infrastructure projects considered necessary to modernize a country through industrialization—they have a particular fondness for dams. To create the initial IBRD fund, each member nation pledged a sum of money proportionate to its national income, which determined the percentage of that country's votes. The US pledge of $3.2 billion earned it the largest share of the votes, 35 percent, with Britain next with 14 percent.[15] In 2012, the United States was still the single largest shareholder with 16.41 percent of the vote, while the forty-seven sub-Saharan African nations commanded less than 6 percent of the votes.[16]

To keep these funds replenished the Bank charges borrowing countries interest on IBRD loans. However, several times during its history, it has had to return to member countries for additional pledges when interest payments have not been sufficient. Approval of these pledges by member countries affords an increased opportunity to apply political pressure on the Bank to either make or cancel loans to countries. A country's influence is greater when it encourages rather than attempts to cancel a loan as the Bank's own ideology is to lend as much as possible.[17]

A second financing mechanism at the Bank provides even greater opportunity to influence Bank policy. The International Development Association was created in 1958 by the Bank to forestall a proposal in the United Nations to set up a separate agency outside of the Bank's control that would provide loans to less creditworthy countries. The IBRD could only lend to countries that were *not* able to get private financing, yet were still judged able to repay their loans, leaving the Bank with too few "appropriate" countries to whom it could lend.[18] The new IDA maintained the Bank's control of development loans and increased the number of countries to whom it could lend. Funds for the IDA are not raised on the bond market like for the IBRD, but are pledged approximately every three years by member countries. As of 2012 the IDA fund has had to be replenished sixteen times. Each of the sixteen times, the process has run into trouble in the US Congress, which has frequently threatened to cut off or reduce funding for IDA. "Congress's most effective weapon against the Bank has been not its directives, but its threats to reduce or cut off funding for IDA. Nowadays, nearly all its contributions are conditional on the adoption of specific reforms in the Bank's approach...."[19] For the sixteenth replenishment for 2011–2014 "donors provided 60 percent of the total $42 billion allocated, with the remainder coming from loan repayments and other arms of the World Bank."[20] While these replenishments do not influence voting shares, donors do place themes or conditions on their donation. In the last two replenishments the donors have agreed to limit the "themes" to climate change, gender, fragile states, and aid effectiveness;[21] however, that does not prevent major players from adding additional stipulations. For instance, the US Congress in 2005 stipulated that 20 percent of the US funds would be withheld unless the Bank withdrew its pursuit of the use of country services in procurement.[22] "Country services" is a policy under discussion at the Bank to allow borrowing countries to have more control over who receives contracts to implement Bank projects in their country. Historically, most of the contracts have been designed by the Bank and awarded to contractors in the global North, frustrating developing countries' ability to control their own economic development. Beyond direct control through the exercise of its large voting share, appointment of the president of the Bank, and general oversight of policies, the United States still successfully engages in brinkmanship to force policy changes every time any of the Bank's funds needs replenishing.

The International Monetary Fund

The United States has been forced to use less direct control over the Bank's sister Bretton Woods institution, the International Monetary Fund (IMF), because the head of the IMF has traditionally been European. That, however, does not mean that the United States has refrained from exercising considerable influence over whom that European would be. In 2000, it virtually blackballed the European Union's initial selection, Ciao Koch-Weser, the state secretary in the German finance ministry. "In vetoing Koch-Weser, the US has made clear that the new IMF chief, whether he comes from Europe or elsewhere, must be completely attuned to the interests and demands of Wall Street and key American banking and finance interests when the next global crisis arises."[23] As with the Bank, influence in the IMF is based on the size of a country's economy. The IMF generates most of its financial resources based on quota subscriptions determined by a country's share of the global economy. The United States, continental Europe, and the United Kingdom have historically had the largest quotas based on their contribution to the world-economy. However, the growth of global South economies has challenged this distribution of quotas and threatened the legitimacy of the IMF. For instance, the contribution of the United States and eurozone + UK in the decade of the 1980s to the world-economy was one-fifth each, with Japan contributing about one-tenth. In the decade of the 2000s the eurozone + UK contribution fell to 6.4 percent from 20.6 percent in the 1980s. The US contribution plummeted from 21.2 percent in the 1980s to 9.6 percent in the decade from 2001 to 2012, yet there have been only modest reflections in the allocation of quota shares. This new imbalance in the size of a state's voting share is posing a threat to the IMF's legitimacy but so far has not been adequately addressed by the IMF.[24]

This system of weighted voting in the Fund and the Bank skews the voting heavily in favor of the developed nations, giving them immense control over the economic and social policies of the developing world, especially the poorest nations. Before the reforms in 2008 and 2010, the developed countries, with less than 14 percent of the world population, had an overall majority of the votes in the IMF (60.4 percent), the IBRD (57 percent), and IDA (60.1 percent). The whole of sub-Saharan Africa had fewer votes in the IMF (4.6 percent) than France (4.9 percent), the UK (4.96 percent), Germany (6.1 percent), or Japan (6.15 percent), and less than one-third held by the United States (17.14 percent).[25] This weighted voting in the IMF and the Bank had given developed countries, with only 14 percent of the world's population, power to control the shape of the entire global economy.

Pressure to reform this weighted voting in the IMF resulted in reforms in 2008 and proposed reforms in 2010 that are projected to go into effect in 2013. These reforms were formulated before the Great Recession, which peaked in the beginning of 2009, and are relatively modest, according to IMF's own board operations chief.[26] The 2008 reform shifted only 2.7 percent of votes to developing countries, and while the IMF claims that the 2010 reforms, if approved of by the United States, could add up to 6 percent shares for dynamic emerging market and developing countries (EMDCs), critics challenge these estimates. They argue that half of this 6 percent will come from "developing" countries and not advanced economies, "which will drop from 57.9 percent to 55.3 percent, a loss of only 2.6 percent."[27] Even with these reforms, the United States still maintains special veto power over decisions where a supermajority of 85 percent of members must approve and the United States has a voting share of 16.75 percent.[28]

In the early part of the 2000s, some political and economic thinkers had wondered if the IMF's time was almost up.[29] But, with the Great Recession, the G20 (the coalition of Western wealthy industrialized countries) needed to stem the volatility of the world market caused by corruption of real estate markets in the United States. In reaction, the G20 agreed to triple the budget of the IMF, and even open up G20 countries to increased surveillance by the IMF in return for monies to bail out troubled economies around the world. The IMF managing director, Christine Lagarde, secured $438 billion as of June 2012 from mostly emerging economies, including China and Brazil, without the help of the United States. With most of this money targeted to assist struggling eurozone countries, the donor countries, especially Brazil, are arguing that the recent 2010 quota reforms, which went into effect in June 2013, were set up before the financial crisis and should be changed to grant greater rights within the IMF to the emerging economies.[30] In June 2013, the Obama administration submitted these reforms as part of the proposed budget. The divided US Congress has blocked these efforts, and as the United States is the only country with veto power in the IMF, they likely will not go into effect any time soon.

The power behind *industria* is even more concentrated when you consider that state power exercised through the international financial institutions (IFIs) is largely circumscribed by TNCs and a few wealthy elite in core and periphery states who maintain control over TNCs.[31] Sklair contends that "global capitalism, driven by the TNCs, organized politically through the transnational capitalist class, and fuelled by the culture-ideology of consumerism, is the most potent force for change in the world today."[32] The sole accountability of the TNCs is to their stockholders who generally become concerned only if their dividend payments decline. Michel Chossudovsky expresses this well when he asserts that the Bretton Woods institutions create an enabling environment for global banks and multinational corporations. These regulatory bodies interface with Wall Street bankers and the heads of the world's largest business conglomerates through organizations such as the Davos World Economic Forum, the Washington-based Institute of International Finance, and the International Chamber of Commerce; in addition, "other 'semi-secret' organizations—which play an important role in shaping the institutions of the New World Order—include the Trilateral Commission, the Bildebergers, and the Council on Foreign Relations."[33]

Development Discourse Normalizes the Power Grab

To understand how this imbalance of power was orchestrated, we have to look at two components. The first is how this ruthless power grab has been justified and, second, how the more modest original mandates of the Bretton Woods institutions were allowed to burgeon into the corporate takeover that we see today. Whenever there is a power grab, those taking power have to justify the acquisition and set up why they should be the ones to have the power. This feat of magic was largely initiated by President Truman in his 1949 inaugural speech, in which he declared that more than half of the world's people were living in misery and poverty, and were thus vulnerable to fall prey to Communism. Truman had invented a new problem, "underdevelopment," and the proposed solution was greater production through "more vigorous application of modern scientific and technical knowledge."[34] As Arturo Escobar in his now classic study of development contends, reality had been colonized by development discourse.[35] From then until today, there would emerge slightly different paths to this thing called "development," but the fact that there are underdeveloped economies that must become developed through processes of industrialization and urbanization, could not be challenged. What also could not be challenged was that the path to development must come from economic growth aided by modern science and technology. The assumptions embedded within identitarian ontology and its corresponding epistemology of progress achieved through modern scientific rationality formed the basis of development discourse.

The key to unmasking the forces behind globalization today is to render visible the power of discourse to set the rules of the game. Who has the power to speak and from what authority?[36] Discourse is not just words or an expression of thought, "it is a practice, with conditions, rules, and historical transformations."[37] Discourse does not rely on force to gain power; instead it normalizes systems of power. Escobar contends that most forms of understanding and representing the Third World do not act "so much by repression but by normalization; not by ignorance but by controlled knowledge; not by humanitarian concern but by the bureaucratization of social action."[38]

In this normalization of power, the key is whom the elite select to have the authority and moral force to set the discourse. For much of modernity the right to "serious speech acts" has been handed over to "experts" trained in Western institutions in fields such as statistics, demographics, engineering, public health, agriculture, and most critically, economics. These experts of modernity have been anointed by the elite with the authority to set up policies designed to modernize the "poor," "backward" peasants and Indigenous peoples of the Third World. In the process, the inhabitants of these "dark" and "dangerous" places were stripped of their cultures, their humanity and reduced to statistics—objects to be manipulated for their own good by the new experts. The "Third World and its peoples exist 'out there' to be known through theories and intervened upon from the outside."[39] They are often characterized as infantile and with problems—backward, uneducated, lactating mothers; underfed children; and disease-prone.[40] This comports perfectly with identitarian ideology of separateness in which every separate part becomes categorized in whatever way those who dominate the discourse want. These people were not "like us"; thus, they had no human connection to those in the developed world that would necessitate recognition of their full personhood.

Development discourse's primary tool of objectification has been and continues to be economics. From the moment Truman identified underdevelopment as the problem, economics became the only recognized cure. Underdevelopment was interpreted as poverty that resulted from the lack of income. Income was to be raised in the Third World by the same path taken by developed states: They must industrialize, so the argument went. There was no recognition that the path taken by the industrialized countries had depleted their own resources and polluted their environment to reach their level of development. Once the problem was identified as a lack of income, the only aspect of development that could be considered was increasing the income per capita, which reduces people to statistics rather than considering their total way of being in the world.

With the elements of discourse in place all that was needed was to translate these values into actions while keeping the pea thoroughly hidden from sight. This sleight of hand was performed by concretizing the discourse into practice by bureaucratizing it, and the ideal bureaucracy for development discourse in the late 1940s and 1950s was the newly set up World Bank. Using the discourse of development economics that stated that economic growth, pursued through rational planning steps, would erase poverty in the Third World, the World Bank began setting itself up as the authority on development practices. By hiring experts who were all immersed in the same discourse, the Bank was able to argue that it was setting up bureaucratic practices based, not on politics or ideology, but rather on sound, rational, value-neutral principles based on science. The only people not "educated" in these principles were the peoples upon whom these practices would be implemented, but, of course, they were not allowed to ask questions because they had been infantilized and stripped of their agency. The experts from the Bank and its entire network were then not held accountable, because their practices were based on rational value-neutral policies that no one immersed in development discourse would challenge, including elected representatives in the developed states. As Escobar contends, the World Bank became expert at the god trick of seeing everything from nowhere, in which they transform conditions of people's lives into "a productive, normalized social environment; in short to create modernity."[41] In sum, the power grab of the Bank and its most powerful member states was made invisible by development discourse, which normalized its practices as the only viable solutions to the problem of poverty. These backward, child-like peoples needed to become civilized or they would fall prey to Communism.

Bretton Woods Institutions Expand Their Power

With the exercise of power by the Western states disguised as rational bureaucratic practices of the Bank, the next stage was set for the mandate of the Bank, and then the IMF, to burgeon into the economic globalization we see today. The first step in this process was a stipulation of the Bank's operations stating that "it would lend only foreign exchange and it would insist on being repaid in the currencies it lent. Therefore the projects it funded would have to earn foreign currency."[42] The government of a country would not only have to guarantee the loan but the project would also have to be designed to produce exports that could earn foreign currency with which to pay off its loans.

This meant that, even though the declared goal of the Bank was to lend Third World countries monies to develop their own economies as they saw fit, the actual projects selected would have restrictions on them from the beginning. The most fundamental restriction was that funded projects must lead to the development of products for export that could earn foreign currency. The primary exports developed were raw materials and agricultural products, which had very little value added. Moreover, these products must be something that met the needs of the core processes in the industrialized world. Escobar identifies five imperatives the United States faced after World War II—"to consolidate the core, find higher rates of profit abroad, secure control of raw materials, expand overseas markets for American products, and deploy a system of military tutelage."[43] The implementation of these priorities will lead to the final structure of *industria* today in which we have mature economies at the core of the state-corporate network that are supplied with raw materials and cheap labor from the periphery, whose countries remain permanently stuck in a dependent economy with little capacity to industrialize.

The first expansion of the Bank's mission came as it moved from rebuilding war-torn countries to pursue "development" in the Third World. When President Truman saw the Bank was moving too slowly, he set up the Marshall Plan for Europe. The Bank then had to cast about for a new mission and turned its gaze to the countries of Latin America, Africa, and Asia. Bank policies prohibited it from formally interfering with the internal affairs of a country, but Bank presidents continually ignored this proscription, propelled by the belief embedded in development discourse that these primitive countries needed Western expertise to properly develop the industrial base required for modernization. By 1949, Eugene Black was the Bank's third president and the first more-than-transient leader. The first president had only lasted six months and the second only hung on for two embattled years.[44]

Black served for thirteen years during which time he set in motion the Bank's procedures for loan applications. Immersed in the newly forming development discourse, he made it clear that the Bank would refuse to lend money to countries that he considered to be pursuing "unsound policies."[45] In order for a country's loan to be approved, it must draw up comprehensive plans for how it intended to develop. When Colombia in 1949 agreed to develop such a plan, it asked the Bank for assistance. Black was only too pleased to comply and set up what became known as the Currie Commission, named after the head of the commission. The Currie Commission was a fourteen-man team comprised of "Bank staffers, U.S. government employees, and independent consultants, all but two of whom were American citizens."[46] To disguise the Bank's role, the Colombian government agreed to set up a committee of prominent Colombian citizens to formally endorse the comprehensive plan for future development in Colombia. Even with this rubber stamp of a "national plan" in place, the Bank, still doubtful of Colombia's expertise to actually implement the plan, suggested that they establish a National Planning Council to oversee its new development program. Colombia agreed, and Currie and an American economist were installed as advisors to the council making the Bank the "driving force behind the decisions Colombia was making about its long-term development."[47] The Bank generously responded with more than $300 million in loans to Colombia between 1956 and 1963, two-thirds of which went to autonomous public agencies it had helped to set up, not the central government.[48] This model of setting up autonomous agencies, separate from the central government and often outside of legislative control of the country, became so dominant at the Bank that by the "early 1970s, more than half of all its loans went to autonomous agencies it had helped to establish in scores of countries."[49]

The arrogance exhibited by Bank experts in the 1950s was based on a disdain for local knowledge and opinion that still dominates Bank mentality today. "It was an article of faith to Bank staffers that only highly educated specialists had the skills and the knowledge needed to guide developing countries into the future."[50] Black himself asserted that most people in developing countries would not abandon old habits and the "apostles of a new life" would only emerge after "close contact with Western education, Western political thought, and Western material living standards."[51] To aid their enlightenment, the Bank set up its own instruments of indoctrination, beginning with the Economic Development Institute (EDI). With funding from the Ford and Rockefeller foundations, the EDI taught a selected elite from the Third World how to develop and implement projects suitable to the Bank. These "reformed" elite from the Third World countries, fully immersed in development discourse, would close the circle as they returned to their countries and became the Bank's "local" experts. The entrenchment of development planning was furthered within countries as these new development experts, trained by the EDI and other Western institutions, replicated the new institutions of planning at the level of cities, departments, towns, and rural areas. "Through this network of power, the 'poor,' the 'underdeveloped,' the 'malnourished,' and the 'illiterate' were brought into the domain of development."[52]

With reality now firmly colonized by development discourse, the way was clear for the Bank to reject Indigenous plans for development and promote its favored giant infrastructure projects—dams, railroads, airports, and ports. These were seen as critical to industrialization—the sole path to economic growth for the Bank. They also were seen as the most likely to produce profit in foreign exchange in the shortest time horizon possible by building the infrastructure required to facilitate export of natural resources to core countries. Moreover, they would also satisfy its Western members who thoroughly expected to get the contracts to build these projects. Fifty-six percent of all the money the Bank lends is paid to companies located outside of the developing country, mainly those from rich countries. The "Bank's rich members expect to get at least as much out of the institutions—in the form of contracts—as they contribute to it."[53] When Colombia in 1952 requested financing for a water treatment plant, the Bank rejected it as social spending inappropriate for a bank to be concerned with. Education and other social projects were discouraged by the Bank, even if developing countries used their own money.[54] The only projects considered were ones that *industria* supported.

As the thirteen years of Black's reign came to a close in 1963, the Bank had lent more than $7 billion for fostering more rapid economic growth of medium- and large-scale manufacturing and commercialized agriculture. Two billion of that amount was for dam building alone.[55] Dams were the Bank's favorite infrastructure project. Not only did they provide electricity essential to industrial growth, they also provided water for the new commercialized agriculture that came to be known as the Green Revolution, which will be dealt with at length in Chapter 4. What the Bank was not looking at were the environmental and social costs of these projects. These were to remain the hidden costs of production that, with the compliance of states, would be externalized onto society.

The power grab by the Bank was cloaked in the moral authority bestowed by development discourse that imbued the Bank, and its fellow Bretton Woods institutions, with expertise steeped in modern scientific and technical knowledge. They were considered the indispensable path to bring these backward, underdeveloped nations into civilization and progress symbolized by Western values. The Bank's stipulation that all loans must be repaid in foreign exchange opened the door for the Bank to dictate the internal economic structure and, as we will argue, the social structures of the Third World. Once the door was open, development discourse was bureaucratized into practices through its newly educated development planners.

The only measure of success the Bank entertained was economic growth, measured as gross national product (GNP) in the developing world. And the developing world's GNP did increase by a rate of 4.8 percent a year between 1960 and 1967, which was faster than in both the United States and Europe.[56] However, locked in its hubris, the Bank refused to see warning signs that it was facilitating a world debt crisis as had happened in the 1920s and 1930s. As countries became too heavily indebted to make their payments, the Bank created another type of loan, a balance-of-payment loan that was quick disbursing. Not only did the Bank increase its lending at a dizzying pace for the next twenty years, it encouraged private banks to join in. By 1970 private banks supplanted the Bank in supplying foreign capital to the Third World.[57] In 1974 private banks, flush with petrodollars from the Organization of Petroleum Exporting Countries (OPEC) oil embargo, increased lending to Third World countries, confident that the Bank and IMF would bail them out because they had followed the Bank's model of the loans being backed by the government of the country. As the external debt of developing countries increased from less than $100 billion in 1972 to more than $600 billion in 1981 the only change in Bank policy was to lend more money. "It did not alarm the bankers that virtually none of the borrowing countries even pretended that they planned to pay off their debts."[58]

The Third World Is Set on Fire and Neoliberalism Moves to Center Stage

With Third World debt mounting throughout the 1970s, additional economic factors in the emerging state-corporate network of *industria* were converging to insure countries in the periphery would be unable to pay off their debt. The sharp rise in oil prices put inflationary pressures on the world-economy, especially in the United States. In response, the United States began to raise interest rates, which immediately raised the debt service costs in the Third World. Higher interest rates also slowed down economic growth within the countries at the same time that the prices of their commodity exports fell. As Amory Starr explains, "international interest rates skyrocketed and export markets collapsed, so that third world countries that had been making their payments got less for their exports and were faced with higher interest rates."[59] Slower growth and loss of foreign currency from exports combined with higher prices for oil and imported goods from industrialized countries only exacerbated the balance-of-payment problems. No amount of loans from the Bretton Woods institutions or private banks could keep pace with the exploding debt crisis, and on August 12, 1982, Mexico announced that it would not be able to service its outstanding debt. By October 1983, twenty-seven countries, owing a total of $239 billion, had rescheduled their debt to banks or were in the process of doing so. Much of the debt was held by eight of the largest US banks, constituting approximately "147 percent of their capital and reserves at the time."[60] Several of the world's largest banks had more debt than capital and were facing loan default and failure. The World Bank, through its own lending and by encouraging private banks to join in, had set the Third World on fire. It was clear that the stakes were too high for the private banks, especially in the United States, to allow the Third World countries to default on their debt. The firefighters called in to put out this blaze were the very arsonists who had set the fire: the World Bank and the IMF.

The conditions imposed on Third World countries to reschedule their debt made the people of these countries pay a very steep price and constituted the next major structural component that transformed the world-economy into the corporate-led globalization that we characterize as *industria*. The transformation of GATT into the World Trade Organization in 1995 completed *industria*'s structure. But in the early 1980s, it was payback time for industrialized countries, especially the United States. The debt crisis of the early 1980s had erupted against a background of Third World struggles to harness the power of the United Nations as a force to counteract the control of the Bretton Woods institutions over their development. The movement began growing in the 1960s when various Third World organizations and programs had formed, often drawing on the work of Raúl Prebisch, an Argentine economist. "Prebisch's theory centered on the worsening terms of trade between industrialized and non-industrialized countries, which resulted in the South needing to use more and more of its raw materials and agricultural products to purchase fewer and fewer of the North's manufactured products."[61] This "structuralism" that created a new post-colonial version of core and periphery processes was a target of Third World organization including the Group of 77, OPEC, and the United Nations Conference on Trade and Development (UNCTAD). In 1964, UNCTAD became the "principle vehicle used by the Third World countries in their effort to restructure the world economy."[62] With Prebisch as its first secretary, UNCTAD advocated fighting this dependency through stabilization of primary commodity prices and tariff rates to reduce the impacts of trying to trade bananas for refrigerators. Because of the threat of Communism, the United States had curtailed all-out efforts to block these measures, until the debt crisis. "While the United States upheld private enterprise and demanded access for its corporations, it was more tolerant when it came to protectionism, investment controls, and a strong role for government in managing the economy."⁶³ The United States had generally tolerated state policies in the Third World that aimed at protecting key sectors of their economies from foreign competition. It had limited its actions to strategies of increasing the powers of the Bretton Woods institutions; delay, in which it would agree to trade policies, but simply not implement them; and cooptation of Third World elites. However, fueled by critiques from the emerging American political right, the UN system began being cast as an enemy of the United States when Third World nations used it to increase control over their natural resources and economy.

When Ronald Reagan assumed the presidency in 1981, he joined Margaret Thatcher to take up the mantle of the right to institute a total *neo*liberal agenda. At a 1981 North-South conference in Cancun, Reagan delivered his "magic of the market speech" to announce the end of development economics and the institution of neoliberal policies of the free market as the only solution to poverty. The debt crisis of 1982 gave Reagan the excuse he needed to strike back at Third World resistance claiming a mandate to not only roll back Communism but also discipline the Third World nations. "What unfolded over the next four years was a two-pronged strategy aimed, on the one hand, at dismantling the system of 'state-assisted capitalism' that was seen as the domestic base for southern national capitalist elites and, on the other, at drastically weakening the UN system as a forum and instrument for the South's economic agenda."⁶⁴

Starr defines neoliberalism as "the political discourse/ideology that recommends deregulation, privatization, and dismantling of the social contract."⁶⁵ Escobar characterized the shift to a neoliberal agenda as "statist and redistributive approaches gave way to the liberalization of trade and investment regimes, the privatization of state-owned enterprises, and policies of restructuring and stabilizing under the control of the ominous International Monetary Fund (IMF)."⁶⁶ The structural adjustment programs (SAPs) of the IMF and the Bank embodied this neoliberal ideology of an unrestricted market as the only solution to economic growth.

The IMF had already had a chance to observe neoliberal economic restructuring programs in the CIA-led coup d'état in Chile in which General Augusto Pinochet overthrew the democratically elected government of President Salvador Allende in 1973. Chossudovsky, while teaching at the Institute of Economics in Chile, got an up close view of his fellow economists who had been trained in Chicago under the tutelage of Milton Friedman as they orchestrated the first economic shock treatment. Virtually overnight, the price of bread increased 264 percent as wages were frozen. "From one day to the next, an entire country was precipitated into abysmal poverty: in less than a year the price of bread in Chile increased thirty-six times and eighty-five percent of the Chilean population had been driven below the poverty line."⁶⁷ Another military takeover, accompanied by an imposed set of "free market reforms," occurred in 1976 in Argentina with similar results to Chile. Chossudovsky contends that these neoliberal assaults on national economies were simply a dress rehearsal for the far more devastating SAPs instituted by the IMF and the Bank.

Since World War II, the IMF had limited its activity to regulating currencies until the world shifted away from the gold standard in the 1970s; it then drew on one of its original unused mandates to manage balance-of-payment difficulties, which increased its role in the national policies of the developing world. Drawing on the monies from the quota subscriptions, the IMF provided a loan of foreign currency in return for structural policy changes made by the country. These SAPs constituted the most significant increase in the IMF's power to dictate the internal economic and social structures within Third World countries. The IMF was now in the development business. The Bank also had been using balance-of payment loans since 1979, but greatly increased the percentage of this type of loan in response to the debt crisis of 1982. By 1986, the Bank finalized its own neoliberal conditionalities with its structural adjustment loans (SALs). These were more of a formality in reality because the Bank still predicated its lending on a country's compliance with all IMF conditions. The impact of virtually all of these imposed conditions fell disproportionately on the poor and the environment. Chossudovsky asserts that this New World Order "feeds on human poverty and the destruction of the natural environment. It generates social apartheid, encourages racism and ethnic strife, undermines the rights of women and often precipitates countries into destructive confrontations between nationalities."[68]

The conditionalities are designed to replace a government's role in the economy with the private sector, which most often means transnational corporations replace the government because small and medium-sized manufacturing, agriculture, and business concerns are unable to withstand the competition and soon collapse. With drastically reduced government social spending in place, the country is then theoretically free to use the revenues saved to repay external debt. The IMF conditionalities were originally considered as temporary economic shocks designed to stabilize an economy in severe balance-of-payment deficits, inflation, and related problems surrounding the debt crisis. As the "temporary" shocks failed and economies crashed, anti-IMF riots broke out all over the Third World. In Tunis in 1984, bread riots erupted as food prices soared. In 1989, food riots broke out in Caracas, sparked by a 200 percent increase in the price of bread. This led to a state of emergency in which the regular units of the military fired on men, women, and children indiscriminately. Other countries boiled over with anti-IMF riots including student riots in Nigeria, which closed six of the country's universities.[69] In reaction to the riots the Bank stepped in with a second generation of structural adjustment loans, which were supposed to be less harsh than the IMF conditionalities, but countries still had to be in compliance with IMF conditions to receive Bank loans of any kind. Additional threats to countries that did not comply with IMF conditions included restriction of access to IMF loans; restrictions to other creditor countries, which generally require the countries to be in good standing with IMF requirements, meaning that a country may not be able to reschedule its debts; and finally, potential investors may withhold support from a country out of compliance with IMF conditions.[70]

Structural adjustment programs of the Bank and IMF continued to fail, leading to further debt crises in 1997, and then in the Mexican and Asian economies in 1998. Once again, the Bretton Woods development discourse placed the failure of their programs at the feet of Third World "incompetence, inefficiency, corruption and cronyism. And the bailouts by the IMF—in the end by taxpayers—were made to seem like massively beneficent acts of charity on our part toward these dysfunctional underprivileged Asian and Mexican friends."[71] Jerry Mander continues by contending that bailouts really go to the American, European, and Japanese financial institutions that largely caused the problems to begin with by "advocating and then stimulating overspecialization, overexpansion and export-oriented production within nontraditional economic areas."[72] He points out that "taxpayers entered the breach to pay back bankers for their own horrible mistakes, but left the victimized countries in very rough waters. Banks were rescued, not countries."[73] This pattern was repeated in the United States with the bailout of banks "too big to fail" but not main street businesses or citizens' mortgages in the financial crisis of 2008–2009.

Development discourse, now locked into the neoliberal form of colonization of the Third World, maintains that development economics only failed to industrialize countries because the plans were not fully implemented or because the countries' own incompetency and corruption led to the failure. While there is some partial truth in these claims, the fact remains that the Bank had set up parallel governments within countries to implement the policies the Bank itself had designed. Civil society and often even the county's own elected legislatures were not allowed to give input to these grand designs. However, with the blame fully placed on the failure of Third World governments' oversight of their economy, the way remained clear to dismantle government involvement and turn virtually all of the functions of government over to the private sector, which neoliberal dogma always asserts is more efficient. Petras and Veltmeyer unmask this deception by pointing out that these public services provided by governments, particularly in Latin America, were the result of public demand for social services backed by private industry's need for public infrastructure. These popular struggles for "drinkable water, adequate and inexpensive transportation and investment in strategic electrical and energy sectors led to public demands for state intervention to prevent health epidemics and provide infrastructure to facilitate trade and manufacturing."[74] Labor and capital both supported the government's involvement in the economy until labor began to organize for increased social services. Capital, having benefited from government support, was now more powerful and had a new agenda. Employers "after securing state protection, subsidies and monopoly prices sought lower labour costs and greater freedom from state and labour obligations to increase profits and diversify their investments."[75] The authors contend that it was not so much a "profit squeeze," as employers wanted to "dispose of profits, capital and investment whenever and wherever they wished."[76] In other words, the new neoliberal economics would allow capital and investment to globalize, but restrict labor movements to a weakened position with only castrated states left to protect their interests in the face of corporate-led globalization. "Basically the choice was either to support peasants against landlords, who in many cases included industrialists and their immediate families and financial associates, or seek to appropriate public enterprises and a greater percentage of the state budget and go 'overseas.'"[77] Large-scale accumulation from high profits, accrued under protectionist regimes, was channeled overseas in the form of portfolio investments, and overseas partnerships, which could access technology, markets, and financial resources.[78]

Bretton Woods Institutions Reject Criticism and Refuse to Share Power with Civil Society

When examining the impacts of the Bretton Woods institution's development programs over the years, what seems to virtually leap off the pages of their history is a cavalier attitude toward the actual effects their policies have on the people impacted. Since Truman, their discourse has always professed a concern for alleviating poverty, yet the poor are the most negatively impacted and there is a growing gap between the wealthy and the poor. A report from the highly respected international aid agency, Oxfam International, expresses this conundrum: "It seems impossible that the World Bank and International Monetary Fund (IMF) would give advice to developing countries without fully considering how it might affect the lives of poor people."[79] Yet, as Oxfam points out, the Bretton Woods institutions' own internal documents continue to reveal a hubris that facilitates a rejection of local expertise and opinion. Despite calls in 2005 from G8 leaders meeting in Gleneagles and an OECD Paris Declaration on Aid Effectiveness stressing the importance of participation from those most likely to be affected, neither institution is supporting country-led development plans, much less ones that genuinely take the plight of the poor into account.[80] With the US veto of the Bank's attempt to introduce "country services" to allow some country-led development, it seems as if Third World countries will continue to be unable to control their fate.

The only in-depth examination the Bank has participated in was one begun in 1996 in a tripartite structure consisting of the Bank, national governments, and SAPRIN. SAPRIN, the Structural Adjustment Participation Review Institution Network, was a global network of citizen groups established to expand and legitimize the role of civil society's input into economic policies imposed by SAPs. "In short, the IMF, the World Bank and their most powerful board members still determine national economic frameworks, while the majority of the populations, particularly the poor, as well as important economic sectors, continue to suffer the effects of these policies."[81] The project got under way in 1998 with the final selection of nine countries—Ecuador, El Salvador, Bangladesh, Ghana, Mali, Uganda, Zimbabwe, in addition to Mexico and the Philippines—whose civil society organizations continued to participate even when their governments dropped out. El Salvador quit halfway through, leaving eight remaining countries. At the heart of the five-year process were two national public fora in each country in which members of SAPRIN, the Bank, and governments heard directly from citizens and civil society organizations who had been impacted by SAPs during the 1980s and early 1990s. Peoples' voices were no longer to be silenced. Prior to the first forum, technical teams from each organization did field research in all ten countries to narrow down the issues to three or four. All members of the tripartite structure agreed to the initial reports, but at the very first forum, the Bank began to reveal resistance to being held accountable by the people. "The Fora were designed as a learning experience for the Bank and governments, but the sessions invariably turned into debates, with the Bank defending its record and disputing citizen analysis."[82] The next step was to take the information provided by the citizens and have each country's teams conduct research on the issues raised. The Bank balked again, insisting that it be able to use the same "local experts" it had trained, but SAPRIN held its ground, insisting on genuine local voices. Often holding workshops on methodology, the research process was designed to "take into account the institutions, power structures and interests that affect economic behavior, and seek a multidimensional analysis that considered factors such as gender, ethnic, cultural and age differences, as well as variations in the political, social and environmental context."[83] These reports were then to be presented in the second public forum. The Bank publicly accepted the reports and promptly withdrew from the project with no acknowledgment that it intended to incorporate any of the "vast amount of potential learning into internal decision-making processes at headquarters."[84] The Bank returned to its god trick of seeing everything from nowhere and accepting no accountability, which has always been how it responds to criticism.

The completed report, officially released in April 2002, found that the "greatest emphasis was placed in all the countries on the problems, dislocations and increased poverty and inequities generated by economic adjustment policies through the destruction of domestic productive capacity and through the failure to generate sufficient employment at a living wage."[85] Listed behind this primary loss of domestic productive capacity were the more specific negative impacts of trade liberalization, privatization, agricultural reform, and impact on workers of labor market reforms.[86]

The SAPRIN report firmly rejected the neoliberal assumption that the private sector is more efficient. Neoliberal doctrine that the private sector is always more efficient in delivery of services is "rooted in the belief that all forms of government spending, including social spending, constitute subsidies, with the assumption that subsidies represent state intervention that interferes in market functioning and leads to economic inefficiency."[87] Some of the worst impacts of these cuts in government social spending were felt in the areas of health and education, in which SAPs almost always require government cutbacks. The SAPRIN report notes that in "every country studied, inadequate infrastructure and the lack of educational materials and supplies as well as declining salaries and insufficient training for teachers, have negatively affected the quality of education provided."[88] With fees replacing public support for education, the poorest parents often had to make a decision on which children could go to school. The children left behind were most often girls. In Zimbabwe the reason "it is too expensive" for not being in school was "more frequently cited in all age groups by girls than boys."[89] The same preference was given to boys in Uganda. Women also were found to be less likely to access medical care because of lack of funds. SAPRIN concludes that education and healthcare under SAPs have been reduced to commodities to be bought in the market by those who can afford them, and, further argues that under adjustments there is a transferring of resources from the poor to the rich as subsidies continue to be "extended to private corporations through credit guarantees, tax incentives and even bail-out packages and loans to rescue ailing banks and corporations."[90]

Additional SAP conditionalities designed to shrink government included cutting government subsidies for basic goods such as bread and energy, which as we have seen, often have led to civil unrest. But SAPs also require governments to rewrite laws to deregulate areas such as labor and environmental protection. Although international financial institutions (IFIs) "are, by statute, prohibited from interfering in the political affairs of their member countries, governments have been pressured to change laws to comply with adjustment conditionality, legislatures have been kept in the dark about loan requirements, and cabinet ministries other than those responsible for strictly financial and economic matters have often been left out of the loop."[91] What labor and environmental laws are left on the books are often not enforced because of cutbacks in government workers. Further restrictions on government functions include privatization of public utilities, including such essential services as the provision of water and electricity.[92] This privatization usually results in higher prices, which affect the poor disproportionately. "In Hungary, for example, rates for electricity, heat, water and gas increased on average twice as much as wages in order to guarantee returns to private investors. Foreign companies took over the oil and gas industry and raised electricity rates until they were ten times higher in 1998 than in 1989."[93]

Shrinking the role of government in the economy is exacerbated by the often devastating effects of trade liberalization, which include eliminating tariffs designed to protect infant industries, opening markets to global competition, and forcing a country to switch to an export-led strategy for development in order to secure foreign exchange with which to service external debt. "The principle instrument of trade liberalization was and is the rationalization of the import regime, which contains measures such as elimination of import quotas, the reduction and unification of tariffs, and removal of special tariff concessions and exemptions."[94] Neoliberal theory contends that the parts of the domestic economy that survive international competition will be more efficient and facilitate specialization. The reality has been very different from the theory. In Mexico the "increase in exports displaced national production for the domestic market, as liberalization policies were not accompanied by industrial or credit policies to modernize production, level the playing field or otherwise provide support for the large number of micro, small and medium-scale enterprises that had made up a large part of the country's product apparatus."[95] The SAPRIN report concludes that overall, "import liberalization reduced or wiped out certain domestic manufacturing activities and forced many small and medium-scale enterprises out of business."[96] More often than not the domestic manufacturing capacity was replaced by foreign-owned companies producing for export in sweatshops and export enclaves. Sklair contends that intense pressure to keep labor costs down for TNCs often leads to creation of export-oriented zones where the rights of workers to organize are either suppressed or curtailed.[97] By the 1990s, more and more of the Third World was being turned into free zones of various types as whole countries began to dismantle their restrictions on foreign direct investment. In export-oriented zones around the world the preferred employees are young, unmarried women who are considered more docile workers.[98]

The SAPRIN report found that SAP export-led strategy consistently imposed on governments showed an increase in exports, but "this export growth was very narrowly based on a few national resources and items produced with low-skill labour. Moreover, export growth was underpinned by continued import growth and falling terms of trade."[99] In Ecuador, there was a 40 percent drop in manufactured exports while exports in primary commodities (bananas and cacao) increased at an annual average rate of 2.7 percent. In Zimbabwe, agro-processing and resource-based manufactured exports such as metal products, leather, hides, wood, and furniture improved while traditional and high-technology industries emerged as low-growth subsectors.[100] "Trade liberalization seems to have unleashed a structural shift away from manufacturing in favour of the service sector."[101] Finally, in many countries "the benefits of export growth went to transnational corporations at the cost of domestic producers."[102]

SAPRIN is a report of the impacts of SAPs during the 1980s and early 1990s, but little has changed. The IMF's name for SAPs is now Poverty Reduction Strategy Facility (PRSF), and at the Bank they are called Poverty Reduction Strategy Papers (PRSP). These are supposed to include Poverty and Social Impact Analyses to predict the consequences of intervention on the various social groups, but as Oxfam International reported, "neither institution is either adequately supporting developing countries to undertake this process, or carrying out the vital analysis themselves."[103] As Juan Zalduendo of the IMF's policy development and review department declared, "IMF conditionality is here to stay."[104] In a study published in 2012, conditionality, or demanding that a country change its spending and economic rules to gain access to IMF and World Bank resources, was shown to cause government crises—especially in the case of the World Bank—in part because these conditions can harm the short-term welfare of citizens in these countries who have little say over accepting the socially dangerous conditions neoliberalism can create.[105] Further empirical research shows that IMF conditions and decisions "faithfully reflect US interests" much less than the actual needs of low- and middle-income countries.[106] Even during the 2008 Great Recession, lending by the IMF was based more on whether a country was allied with the G7 (the United States, France, Germany, Italy, Japan, the United Kingdom, and Canada) rather than if it was in desperate trouble, putting many low- and middle-income countries into extremely vulnerable and unstable circumstances.[107]

Ironically, the initial rules of the IMF had put strict capital controls on countries, but in 1971 the United States broke from these limits, which had required the backing of actual gold for every dollar the government printed. This is why some scholars refer to the current era as the "post–Bretton Woods period" because the gold standard limits of the accord collapsed. The so-called collapse of the Bretton Woods system "lost an institutionalized mechanism by which governments coordinated their joint control over international capital markets during the 1950s and 1960."[108] This initial post-war period fostered Keynesian economic policies that used the nation-state as a counterbalance to soften negative effects of the market, through more government spending in down economic times. However, in the 1970s, the Western world suffered stagnating economies with deepening unemployment and increasing inflation. Coincident with the rise of the modern right wing in the contemporary conservative movement, "policy makers increasingly adopted the view that government interference was the main culprit and that the solution involved reforming the economy in ways that privileged markets' economic influence over that of the state," which crystallized into a perceived agreement called the Washington Consensus emerging from the New World Order.[109] Deregulation, freeing markets from state controls, reducing government spending in social welfare, privatization, and protecting private property were all parts of this consensus among core elites that came to be known as neoliberalism and were globalized through the Bretton Woods regime. As deregulation increased, giant financial interests made incredibly large bets, only to be bailed out by state financing and credit (such as in Mexico in the 1980s) when these bets failed. The international financial institutions (IFIs) would then usually impose austerity (reductions in social welfare spending) programs to pay for the private capital losses. Centeno and Cohen remark, "The pattern was set for years to come: Deregulation would lead to crisis, public authority and money would be used to resolve it, and austerity would be demanded as a way to pay for the mistakes."[110]

This pattern extends to the 2008 Great Recession, which started with a deregulated financial market in the United States that bundled together a gossamer web of tangled ghost-like, supposedly low-risk assets, creating a burst of cash in the system on values that did not really exist. Within months of financial collapse, $50 trillion evaporated from the base of these banks, prompting government to bailout the banks while these very banks evicted people from their homes. Thus, the nation-state remains relevant and necessary under this brand of hyper-aggressive cowboy capitalism because the private risks that firms take are covered by a de facto insurance paid for by the public. As one might expect, it is not a nameless set of stooges who keep this increasingly Manichean and corrupt system going, but it continues as a "product of direct political influence, not just by the amorphous capital markets but by the direct collusion and influence of a narrow group of individuals whose personal or institutional control of vast amounts of money allowed them to buy their respective polities."[111] In other words, this system is driven by a specific set of elites who work at the heights of a coordinated corporate-state alliance and who enjoy "the most definitive gains."[112]

Since the Great Recession, the only loans offered by the IMF that do not have structural conditionalities attached are the so-called flexible credit lines.[113] However, these loans are only given to countries with very strong economic fundamentals and policy track records if they are faced with balance of payment pressures. In other words, their economies already comply with neoliberal free trade policies. Even more critically, most of the trade liberalization policies are now part of the World Trade Organization, which will be examined in Chapter 3.

The Bretton Woods institutions have made it clear that they will not engage in a genuine effort to become accountable for the harms inflicted on the poorest peoples of the world or the devastation of the planet Earth. As Sklair points out, it is the role of the IFIs to facilitate and subsidize the entry of TNCs into the Third World.[114] The Bank fell right in line with the neoliberal geoculture presented by Reagan and Thatcher, which it made official in its 1991 *World Development Report* endorsing a "market friendly approach" to development.[115] The Bretton Woods institutions are bureaucracies "acting on behalf of powerful economic and financial interests. Wall Street bankers and the heads of the world's largest business conglomerates are indelibly behind these global institutions."[116] The power exercised by IFIs is rendered invisible by passing it through bureaucratic practices as argued by Escobar. As explained in the following chapters, states serve corporations by insuring their access to raw materials. States then disguise their role by using discourse to normalize their support for corporations. As Hipwell contends, "Not only do state governments frequently put the interest of corporations ahead of human rights, they have also actively cooperated in repressing civilian opposition to resource exploitation."[117] Key to global corporations maintaining their power to shape the role of states and IFIs is the ability to remain invisible. As Starr phrases it, "if invisibility is crucial both to the system's legitimacy and to its continued relatively unimpeded operations, then discourse can bring 'sunshine' to the workings of corporate hegemony."[118] Her solution is to "name the enemy." The enemy is the corporate structure that benefits from a twofold legal character: first, their supposed service to "common good" trumps any accused social costs or harms and, second, they can claim rights and protections due to citizens.[119] Their only accountability is to shareholders, most of whom are focused on profits, not whether the corporation is making those profits in a socially responsible manner.

So the questions remain, where is the pea, and whose job is it to find it? We argue that the only way the ravages of *industria* can be halted is if the citizens, especially those within the developed world, first educate themselves on what their states are doing in their name, and then organize to exercise their democratic sovereignty to compel their state governments to rein in the power of global corporations and the international financial institutions that they bend to their service.

CHAPTER THREE
INDUSTRIA ENCLOSES THE GLOBAL COMMONS

> The logic of the global market has justified innumerable violations against our and every other community, whether land tenure issues and the allotment processes or deforestation by Weyerhauser, the destruction of the buffalo herd or the diversion of water affecting our wild rice.
>
> —*Winona LaDuke (Anishinaabeg)*[1]

The first two Bretton Woods international finance institutions (IFIs) were used very effectively to construct *industria*'s power/knowledge system primarily to benefit elites in the developed world centers, although selected elites within Third World countries were co-opted and also prospered. The Bank lent money to developing countries for projects that it had chosen and whose benefits would inure to the elite, including mostly being built by core countries' corporations. When developing countries fell behind in the repayment of this onerous debt, the IMF joined the fray by agreeing to lend them money for debt repayment on the condition that the country instituted neoliberal market reforms variously referred to as "structural adjustment policies" (SAPs) or "conditionalities." By the close of the 1980s, both the Bank and IMF were using debt repayment as the mechanism to impose the neoliberal market policies that *industria* needed to force open Third World markets and gain unrestricted access to their natural resources and labor.

Key to this strategy was destroying countries' ability to provide for their own food security. This destruction of food security is especially harmful to Indigenous communities, many of whom practice environmentally sustainable subsistence farming as a vital part of their cultural identity. The first stage was the Bank's insistence on countries' replacing food grown for domestic consumption with cash crops for export. The second involved how these cash crops were to be grown. The core countries not only wanted the export crops, they also wanted the resource of cheap labor to set up offshore manufacturing where they would not have to comply with social legislation that protected labor and the environment, which often existed in core countries.

The initial problem was how to get the Indigenous peoples and peasants, who were largely self-sufficient, off of their small plots and/or collectively owned land and into city centers where they could be forced to work for low wages. Two primary strategies were employed. First, *industria* continued to support World Bank funding of agriculture in the industrial model, which encouraged "land reform" that broke up small holdings and collectively owned land. Loans were then provided to the remaining large landholders to institute monoculture planting of high-yield varieties of seeds that required massive inputs of fertilizers, pesticides, herbicides, and water. The second strategy to weaken subsistence farming was to force open agricultural markets in the Third World so highly subsidized core country grain could be dumped into countries at prices at which no local farmer could grow the grain without being subsidized by their own government. Neoliberal reforms required countries to cut back tariffs that once protected domestic producers from these artificially low-priced grains. Small producers were soon overwhelmed by the importation of cheap grain from the global North and eventually driven off of their land.

We can see the impacts of this destruction of food security in the 2008 worldwide food crises. Countries, which had been self-sufficient in food production before SAPs forced them to stop growing food for local consumption, fell into crisis when drought and diversion of internationally traded grains into biofuels reduced the availability of grain imports. The European Network on Debt and Development (Eurodad) contends that the roots of the food crisis were planted in the industrial agricultural model and SAPs from the international financial institutions. "Yes, indeed, we have heard about drought and poor harvests; increased demand from growing populations; crops being diverted into biofuel production or speculation. But, there is little talk in official circles about a more fundamental cause; and that is the agricultural model promoted since the mid of the last century and the trade liberalization and structural adjustment policies pushed on poor countries by the World Bank, IMF and others since the 1980s."[2] The UN Food and Agriculture Organization (FAO) warned in October 2012 that the food supply system could collapse at any point with world grain reserves at their lowest point since 1974. Worldwide droughts have stressed this system further, causing the FAO to predict a food crisis in 2013 that could spark widespread riots and bring down governments.[3]

The IMF and World Bank, however, were still relatively blunt instruments for enacting total control that *industria* requires to maintain its power over the world's resources and markets. States still had limited power to use the economic tools of some remaining tariffs, import quotas, and state investment to foster domestic development strategies. Additionally vexing to *industria*'s corporations was that too much intellectual knowledge was held in the global commons. As long as knowledge was held in common and was free to anyone within local communities or Indigenous peoples, it could not be commodified and restricted to profit going only to the elite of *industria*. It required a more refined instrument to complete its domination of state power and local knowledge. To justify this power grab they needed to implement *identitarian* ontology, which views all aspects of reality as separate, unrelated entities that can be categorized for the application of instrumental rationality.[4] Ultimately, the goal is to conceptually reduce every aspect of reality to a commodity, which can then be owned in order to make a profit for the owner. This includes all knowledge held in common by local and Indigenous communities throughout the world, in addition to the provision of services basic to human life such as water and energy. The instrument developed not only needed to control resource use, it also had to have the ability to privatize all parts of the intellectual and physical global commons. Once privatized *industria* could remove the ability of local knowledge to spur innovations within communities it could appropriate this knowledge for their exclusive profit.

Transnational Corporations Find Their Instrument— The World Trade Organization

In 1986 in Punta del Este, Uruguay, the ninth round of multilateral trade negotiations under the General Agreement on Tariffs and Trade (GATT) was launched to accomplish this goal. The Uruguay Round was completed in Marrakesh, Morocco, in 1994 and transformed GATT into the World Trade Organization (WTO)—the perfect instrument for *industria*. To accomplish this task, the largest transnational corporations (TNCs) had come together to draft rules for an institution that could accomplish the final stage of the neoliberal agenda. In 1991, Public Citizen, a nonprofit citizen research group founded to promote government and corporate accountability in the arenas of globalization and trade, received a leaked, secret draft of the new rules that would transform the GATT into the WTO. Lori Wallach and Patrick Woodall write, "The new rules were being written surreptitiously, and under the influence of the world's largest multinational corporations, with five hundred US corporations officially designated as formal government advisors."[5]

As the final component of the post–World War II Bretton Woods institutions, GATT had primarily been limited to liberalization of tariffs on industrial products and was merely a contract amongst member countries with very limited enforcement powers. Under the WTO, enforcement powers were strengthened and new areas were added, including intellectual property rights, services and investment measures, and agriculture. A basic tenant of Indigenous communities is an understanding that knowledge is a communal right open to all in the community, not an individual right. Under intellectual property rights, the knowledge held in common by peoples could be usurped, stolen, and then patented. The once free knowledge, open to all community members, was now "owned" by those who held the patent.

Under the services and investment modalities, private companies could buy up government services such as water rights, education, and healthcare and sell them for whatever the market could bear. Agriculture agreements rigged agricultural trade in favor of core countries and transnational agribusiness corporations, while forcing subsistence farmers off their lands. Finally, the WTO's rules would be legally binding and result in loss of "most-favored-nation" status if a country were expelled from the WTO.

Industria created the WTO to enact the final enclosure of the global commons. Its policies were intended to move beyond trade measures and grasp control of increased mechanisms of state power and local knowledge generation to insure absolute corporate-led domination. The WTO is the ultimate mask on the carnival barkers. It declares that its goal is to end world poverty through free and fair trade, but when we peel back the mask, we see the face of pure corporate greed, unaccountable to the peoples of the Earth or their governments.

This chapter will examine how the corporations constructed the WTO to allow TNCs to work behind closed doors to gain control over "large parts of decision-making, even in developed countries, at the expense of the power of the state or political and social leaders."[6] This corporate-led globalization is occurring increasingly to the detriment "of social, environmental and labor improvements and rising inequality for most of the world."[7] We will break down the mechanisms developed under the WTO that are being used to gain this domination over the resource rich countries of the global South. In Chapter 4, we will focus on the impacts of this domination on small peasant and Indigenous communities and their ability to maintain their own ways of life, which we contend are much more environmentally sustainable than the industrial models advantaged by the rules developed in the WTO. Agriculture, or how people provide for their food security, is embedded in a way of life. We, in the developed core, have lost our connection to this and have much to relearn from those who still enjoy a connection with the Earth. In Chapter 5, we will examine the role of the state in this process and propose a "green theory of the state" that must incorporate an acknowledgment of the role of the natural environment and facilitate participation of citizens in decision making to counteract the power of corporate influence in state policies. Finally, we will present the environmental, cultural, and economic advantages of alternatives to corporate-led globalization. It is not inevitable that the world continue on this path, there are alternatives. Chapters 6 and 7 will open the door to view alternative cosmologies of humans' relationship to the planet Earth.

The first goal of the TNCs in creating the WTO was to open up developing countries' markets by dismantling their barriers against imports of industrial goods, services, and agricultural products. Industrial countries in the global North had had a comparative advantage in these areas since World War II, but their total access to these markets of the global South had been blocked by Third World countries that wanted to develop this capacity for themselves. Often referred to as "dependency,"[8] an imbalance of trade is created when one country can trade high-value-added goods for primary goods. The country with the ability to export manufactured goods can develop a thriving economy. Countries only trading primary goods, such as natural resources or unprocessed agricultural products, are continually at a disadvantage in moving up to develop a capacity to industrialize and tend to remain in a permanent state of dependency. A country exporting bananas will have a difficult time achieving equality with a country exporting refrigerators or cars unless they can protect their infant industries as the core countries did during their own industrialization periods. "Industrialized countries and global corporations demanded global 'free trade' exactly while the terms of trade have been declining for developing countries. As exports increase, revenues actually diminish because the cost of imports for a country escalates faster than prices for its exports, requiring it to double or triple an export (e.g., cotton) to purchase the same import commodity (e.g., tractor)."[9] The point of the new rules was to force open markets in the global South to prevent their achieving an industrial capacity that would allow them to compete with the core countries.

WTO and TNCs' Control of Services

The industrial North is also dominant in services, which make up 60 percent of the global economy and about 70 percent of the US economy.[10] Services include everything from healthcare to banking in addition to all the services normally provided by the public sector such as water, energy, education, and infrastructure. Once turned over to the private sector, citizens are reduced to merely consumers devoid of rights, and only receive the services based on their capacity to pay high enough fees to allow sufficient profit to the private provider.

In order to allow global corporations access to these services the General Agreement on Trade in Services (GATS) provision was added to the WTO. Under this provision all domestic laws and regulations affecting trade in services that have been constructed to maintain these services in the public sector under the control of local governments can be ruled as barriers to trade and become WTO-illegal. "The GATS rules benefit global corporations and investors at the expense of local communities and democratic government authority and put local service providers at a disadvantage."[11] These rules diminish the power of local communities to shape their own economic development including the traditional cultural organizational mechanisms that protect the rights of women and Indigenous peoples. These can be ruled as "non-tariff trade barriers." Furthermore, GATS requires that foreign corporations be treated at least as well as domestic corporations. This "national treatment" rule discourages governments from selecting a local service provider or investor in order to develop a domestic source. GATS "forbids not only domestic policies intended to treat foreign services or providers less favorably, but also laws that may unintentionally have such an effect."[12] This could be used to discriminate against environmental policies that encourage recycling or cleanup of on-site toxic waste that a WTO dispute panel could rule favor local service providers.[13] This provision combined with the "most favored nation" rule, which says that corporations from one WTO country cannot be treated more favorably than corporations from another country,[14] ensures that the "GATS regime reinforces the IMF/World Bank agenda to promote privatization of public services through structural adjustment/poverty reduction requirements and targeted loans."[15] "Practically, what the Most Favored Nation rule means is that if a government contracts out a public service activity or gives a tax break, special regulatory treatment or a subsidy to any single foreign service-sector company or investor, it must extend the same treatment immediately and unconditionally to every interested company from every WTO country."[16] Moreover, all government policies not only directly regulating services but also those that affect services, such as general labor market policies, are included under GATS.

Under this agreement, for-profit providers can come into a country, skim off customers who can afford to pay for the service, and leave those unable to pay without recourse, including for basic services such as energy and water. Clear evidence of this disparity can be seen in the impacts caused by privatization of public utilities and services under SAPs. When public services were privatized in Hungary, for example, "rates for electricity, heat, water and gas increased on average twice as much as wages in order to guarantee returns to private investors."[17]

This disparate impact of privatization, falling most heavily on the poor, was evidenced also in El Salvador where the electricity rates of low-end consumers increased at double the rate of high-income groups. This was especially pronounced for poor communities in rural areas, "which private companies do not see as sufficiently profitable."[18] When basic public services are transferred from the state to private, for-profit companies, basic social services like education and healthcare become merely goods to be sold and bought in the market by those who can afford them. Those hurt most tend to be the poor, especially women "who have to take on greater responsibilities for the education and healthcare needs of the extended family, including care of the sick and elderly."[19] In many cultures of the world today, when families are forced to choose which child will receive limited access to education, most still favor education for boys.[20] Women are also harmed by fee-for-service in healthcare schemes when fees are so high they prevent prenatal care or adequate care for childbirth. These are not merely temporary impacts on isolated people; they tend to have long-term negative consequences to a community as a whole. "The deterioration in health conditions among low-income groups and their inadequate access to quality education translate into underdeveloped potential and erosion of the capabilities of individuals and their communities to build sustainable livelihoods."[21]

An even more stark example of complete privatization of a public service can be witnessed in the case of Bolivia in 1998 when the IMF made a loan contingent on the privatization of all public services. Wallach and Woodall report the case of the Bolivian city Cochabamba, which under additional pressure from the Bank, sold its waterworks to a consortium led by Bechtel subsidiary, Aguas del Tunari. What was left undisclosed in the deal was that the city would be responsible for financing the construction of a huge dam, increasing the cost of their water by 600 percent. As drinking water prices soared citizens took to the streets to be met with tear gas and bullets from their own government. After tremendous effort by the citizens, including injury and loss of life, the city eventually agreed to reverse the concession. In 2002, Bechtel filed a grievance with the World Bank demanding a $25 million payment as compensation for their loss.[22] Bechtel filed the grievance with the International Centre for Settlement of Investment Disputes (ICSID), an "autonomous" institution formed by the World Bank for private investors and corporations to essentially sue countries. In 2008, Bechtel dropped its complaint after enormous international pressure organized by the human rights group, the Democracy Center.[23]

The ICSID is a little-known but powerful force in globalization. It comes into play when two countries sign an investment treaty and the ICSID is identified as the judge for settling disputes. The ICSID is a preferred outlet by powerful countries, whose corporations are looking to invest in weaker countries.[24] Powerful countries in the core nodes of *industria* have major investments in weaker countries and they prefer this outlet over domestic courts of countries like Bolivia. Weaker countries accept this binding commitment when they anticipate needing a loan from the World Bank or need the investment from the source country. The ICSID offers little opportunity for appeal, and is a specific gateway for corporations to discipline peripheral states to serve corporate profit, and transfer public goods to private elites.

GATS and Harmonization

GATS's neoliberal policies designed to limit the ability of governments to control their own social and public service policies fall under a general trend, promoted by global corporations, to replace various national product standards and regulations with one global system.[25] This "harmonization" gained a "significant boost with the establishment of the WTO which explicitly requires harmonization of food and technical standards."[26] Although it would theoretically be possible to use this principle to raise standards, "actual WTO harmonization provisions promote lowering the best existing domestic public-health, food-safety, plant-and-animal-protection and environmental standards around the world."[27] These standards place an upper limit or ceiling on standards, which countries cannot exceed, but fail to put a bottom or floor that all countries must not go below. For instance, the European Union banned US beef that had been treated with hormones. The WTO ruled that this ban was a barrier to trade because it was a higher standard than the highly controversial international Codex standard for food safety, which focuses more on world trade than protecting public health.[28] Codex is considered the ceiling for food safety laws under the WTO Agreement on Sanitary and Phytosanitary measures (SPS) and, therefore, the EU could not go above this standard, even though there is considerable research indicating that these hormones are linked to cancer and premature pubescence in girls.[29] The United States, which brought the suit to the WTO at the behest of the US National Cattleman's Association, imposed a 100 percent tariff on various EU exports when it persisted in pursuing mechanisms to keep the hormone-treated beef out of EU markets. Harmonization as a principle implemented by the WTO often turns out to be a race to the bottom in food and public safety and environmental policies.

This race to the bottom is accelerated by an additional component of the harmonization principle of "national treatment" that states that governments must treat domestic and foreign goods the same, no matter how they are produced. Production and Processing Methods (PPMs) cannot be used by a government to treat goods differently based on the way that they are produced or harvested.[30] These harmonization provisions eviscerate the Precautionary Principle, an internationally recognized theorem of public policy, which is usually "understood to mean that in cases where there is scientific uncertainty, governments have an obligation to take action to avoid harm to public health and safety, or the environment, by seeking out less harmful alternatives."[31] Under PPMs, governments cannot restrict the sale of goods even if they are produced or harvested in ways that are destructive to the environment such as clear-cutting forests, salinizing soils and fresh water supplies to grow shrimp in ponds, or using fishing methods that destroy habitat or kill species such as the nets used to catch tuna that ensnare and kill dolphins. For instance, Mexico used PPMs to challenge US policy that banned the sale of tuna caught by massive nets (purse seines) that are laid over schools of dolphins to encircle tuna swimming below, a practice that has resulted in millions of dolphin deaths.[32] To accommodate Mexico's challenge, the US weakened its protection of dolphins killed by tuna encirclement nets with the passage of the ironically named International Dolphin Protection Act in 1997. Under this law, tuna can carry the "dolphin-safe" label caught by these methods as long as the "official observer" on the tuna trawler, which is a football field length with nets stretching out for miles, did not observe a dolphin in that particular catch.[33] Further difficulties involve how to track which catch was the one in which a dolphin was killed. The prior US law under the Marine Mammals Protection Act had regulated on a country-by-country basis. If a country, like Mexico, permitted these encirclement nets to be used, all tuna coming from that country was banned. Now, for all practical purposes it is left up to the individual producers themselves to identify if a particular catch murdered any dolphins. Due to consumer pressure, the United States has maintained a "no encirclement fishing" standard rather than the "no observed dolphin death" standard, in order for the tuna to carry the "dolphin safe" label. However, there is no longer an embargo against tuna caught with encirclement nets and every administration from Bill Clinton to George W. Bush has attempted to weaken the standard to comply with the 1991 GATT ruling. As of 2011, Mexico is waiting for a WTO ruling against the United States restricting its "dolphin safe label" to ships not using encirclement nets. WTO harmonization has weakened the United States protection of dolphins. Richard Peet challenges the notion that the WTO makes fair judgments based on neutral scientific evidence provided by experts. He maintains that this notion is a "figment of the imagination. Instead, the WTO makes anti-environmental judgments biased by its guiding principles of the 'liberalization' of trade and disguises them as fair, equitable and based in the neutral principles of science."[34]

Tariff Schedules Promote Decimation of Natural Resources

The addition of services and investment[35] measures to WTO is designed to provide *industria* with powerful tools to encroach upon states' ability to protect their citizens and environment from harm. However, an even more direct challenge to a state's ability to protect its environment was embedded in the changes made to how tariffs on primary and value-added goods are scheduled. Under the Uruguay Round, tariff schedules promote exploitation of forests, fisheries, and minerals by placing higher tariffs on manufactured goods than on primary goods. For example, "rough tropical timber comes into the U.S. duty free, but plywood veneered with tropical wood has a tariff of 8%, and almost all furniture above a limiting quota receives a 40% tariff."[36] These incentives encourage a "rip and ship" exploitation of natural resources that has resulted in a substantial increase in the global trade of natural resources without recognition of the harms done to local environments or economies. This increased demand is coming almost exclusively from *industria*'s core countries. "The majority of the world's natural resources (roughly 80%) is consumed by the 20% of the population that live in developed countries."[37] Core countries' demand is encouraging the ripping of natural resources from the mountains, forests, and fisheries of the global South with no concern for the resulting losses of biodiversity, environmental degradation, and increased dependency within Third World countries. These countries are left with little choice but to allow even greater exploitation of their natural resources just to gain small earnings. "The WTO tariff escalation both encourages more intensive exploitation of resources and creates an economic incentive against developing countries establishing manufacturing industries to process raw resources—meaning less earnings for poor countries in addition to increased depletion of natural resources."[38] Prior to the Uruguay Round, there were lower tariffs on finished goods than on raw material, which encouraged local communities to develop the capacity to produce manufactured goods and end the dependent relationship. After the Uruguay Round, the terms of trade under the WTO "limit the authority of developing country governments to constrain the choices of companies operating in their territory," at the same time they demand protection for "rigorous property rights of foreign (generally Western) firms," all of which "lock in the position of Western countries at the top of the world hierarchy of wealth."[39]

WTO Encloses the Final Realms of the Global Commons—Intellectual Property and Agriculture

The neoliberal policies created by the Uruguay Round in services and tariff schedules have proved to cause substantial harm to the power of states, their peoples, and environment. However, two last additional areas added—intellectual property rights, broadened to apply to basic life forms, and agriculture—go even further and impact the ability of Third World countries to provide the basic food necessary for survival. For the balance of this chapter we will focus on the impacts of these last two provisions, and how countries and Indigenous peoples have fought back against this corporate-led industrial agriculture and neoliberal policies privileged by WTO rules.

Trade-Related Aspects of Intellectual Property Rights (TRIPS)

While harmonization policies allow for the weakening of environmental standards, and tariff escalation schedules facilitate "rip and ship" export policies that preclude sustainable resource management, intellectual property rules attack the preservation of biodiversity and erode the ability of a country to provide food security for its people. By altering basic patent laws to include "life forms" in the Trade-Related Aspects of Intellectual Property Rights (TRIPS), WTO rules seek to fundamentally change humanity's relationship to the planet from one that recognizes it as a home open to all who depend on it for survival to just another commodity to be bought and sold in the market place. In India the term *vasudhaiva kutumbkna* means "earth family" or the "community of all beings supported by the earth."[40] This recognition of a continuum between human and nonhuman species, past, present, and future was also expressed by Chief Seattle of the Suqamish tribe in 1848. "This we know: the earth does not belong to man; man belongs to the earth. This we know. All things are connected like the blood which unites our family. All things are connected."[41] This cosmology envisions the planet as a commons, open to all who need her for sustenance, but with the recognition that all interaction must be guided by social relations based on interdependence and cooperation.[42] With the inclusion of TRIPS in the WTO this connection is violated, allowing corporations to privatize basic life forms that had always been regarded as part of the global commons open to all.

Now seeds that had been freely exchanged with neighbors and knowledge of medicinal properties of plant life, accumulated over generations and freely given to all in need, can be stolen by corporations and patented. The benefits of the seeds nurtured by generations of community members have been removed from the commons and enclosed into profit for the holder of the patent. If community members try to continue using their own seeds, they can be sued by the holder of the patent. The enclosure movement in sixteenth-century England was precipitated by the need to increase wool production to feed the growing number of water-powered machines for the processing of wool. Peasants who had historically been able to use the land for food production and grazing were pushed off the land and replaced by the landlords' sheep. Today peasants are also being pushed off the land in order to feed the machinery of *industria*. The current enclosure of the commons is the theft of the "collective assets of the poor." Vandana Shiva contends that "without clean water, fertile soil, and crop and plant genetic diversity economic development will become impossible."[43]

The Global South Resists TRIPS

TRIPS is the first international instrument to require intellectual property protection for living forms. While it has been the goal of the WTO to extend this law over all plants and animals, countries of the global South, which are the source of the planet's biodiversity, have been resisting this attempted takeover by the resource-poor core countries of *industria*.[44] This resistance is based on TRIPS's "abrogation of collective and Indigenous rights formally recognized by the global community in the United Nations."[45] Treating Indigenous knowledge as private property that one person can own is rejected in an official "Indigenous People's Statement" released by organizations in more than thirty countries. "Indigenous knowledge and cultural heritage are collectively and accretionally evolved through generations. Thus, no single person can claim invention or discovery of plants, seeds or other living things."[46] This interpretation runs counter to the Lockean interpretation of intellectual property in which the "innovator is deserving not only of having priority use of the product created, but also a reward (just deserts), if his or her idea has contributed to scientific progress."[47] This privatization of communal knowledge transforms an idea into a "thing" that is separate from the historical and material context that nurtured it and only recognizes the importance of the last marginal addition to the knowledge. What is ignored and left unrecognized are the generations of contributions to this body of knowledge. This is a prime example of how the mechanistic ontology of *industria* places fixed, sedentary boundaries on what was considered to be a continuum of connections by Indigenous peoples. Communal knowledge is transformed into a separate thing to be controlled by this state-corporate alliance.

In what is often referred to as "biopiracy," corporations can engage in the "removal of the organism, whether by literally taking the plant, animal, seed, or germplasm and claiming ownership, or by destroying it" while refusing "to compensate or even acknowledge the original cultivators/custodians of the bioresource."[48] Biotechnology corporations can take a traditional plant or seed, change one gene, and claim that they now own the plant or seed and demand that the original cultivator not use the plant or seed without paying royalties to the newly declared owner.[49] For instance, the US Department of Agriculture and W. R. Grace jointly claimed to have "invented" the use of the neem tree as a natural pesticide and medicine, and succeeded in getting a patent for it from the European Patent Office. The neem has been used in India for more than 2,000 years for these purposes and because of a patent issued in a foreign country, Indians were no longer going to be able to use it for their traditional uses. Through a truly heroic ten-year battle, the patent was eventually revoked,[50] but most Indigenous people and peasant farmers do not have the resources necessary to challenge these cases of biopiracy. Patents are rarely challenged, and generally tend to be granted because patent examiners often do not have the ability to test the alleged "new trait" leaving the only recourse open to the original cultivator to challenge the patent's validity through civil litigation.[51] The fact that more than 97 percent of the patents are held by industrialized countries in the core of *industria*[52] demonstrates quite emphatically who benefits from this process.[53]

This liberal view of knowledge as property is very distinct from views of knowledge derived from exchange and requiring reciprocity. As Mushita and Thompson point out, like love, friendship, and storytelling, knowledge often increases when it is shared.[54] A person who shares seeds with a neighbor gets to see if that seed can be improved by how the neighbor plants and cares for it. If she shares it with even more neighbors, it will only result in further empirical testing of her original seed. Not only does sharing of seeds increase the genetic diversity of seeds, sharing also extends community. "As seed is shared it extends the community, for a seed traveling with a family member or a friend to a more distant place takes with it the taste, texture, foliage, and scents of the previous community. A farmer finding out that a cousin living miles away is using the same maize seed or bean seed invokes pride and a sense of belonging."[55] TRIPS denies this source of community and biodiversity and the "open palm offering seed to share is received by a clenched fist, symbolizing enclosure of the global gene pool."[56] This destruction of biodiversity is particularly devastating on marginal agricultural lands where seeds have been nurtured for generations to withstand drought and other occurring regional stresses. We will focus on the impacts of this threat to biodiversity in southern Africa in the next chapter.

TNCs Use the WTO to Take Over the Global Food Supply

The final element added to the WTO was the Agreement of Agriculture (AoA). In order to get developing countries to agree to TRIPS, GATS, and TRIMs, in addition to lowering their tariffs, the negotiators at the Uruguay Round promised them increased access to the agricultural markets in core countries in the AoA. Without the ability to protect infant industries through tariffs, it became even more critical for periphery countries to have access to core countries' markets for their agricultural products. Under GATT, agricultural products had been allowed special exceptions, including subsidies on domestic production and exports, which favored core countries (especially the United States) whose governments had the ability to provide supports for agriculture. Periphery countries had to rely on tariffs and limited quotas to protect their farmers from the heavily subsidized core country agriculture. The AoA was supposed to guarantee that, if developing countries removed their protections and gave core countries greater access to their nonagricultural markets, core countries would phase out their subsidies on agricultural exports and allow developing countries increased access to their agricultural markets. However, the bargaining power of the developing countries was minimal. Martin Khor, director of the Third World Network (a network that monitors North-South development issues), explains this imbalance of power. "Most of the developing country governments were not active in the Uruguay Round because they didn't have enough officials to participate and they didn't fully understand the issues, so it was not difficult for the rich countries to win them over."[57] Those countries that did try to resist were made to "understand" the interdependence of the WTO and her sister institutions the World Bank and the IMF upon whom many of the poor countries were dependent for loans to pay their debt.[58] The rich countries (the United States and Europe as well as Japan, Canada, and Australia) "have much more negotiating machinery and capacity. They are faster, and are backed up by research institutes and a whole army of bureaucrats."[59]

It was actually no secret that the core countries wanted increased access to Third World markets. Charlene Barshefsky, the US Trade Representative, in 1996 made it clear that the aim of the United States was to gain greater market access in developing countries for American companies.[60] While the core countries did get greater access to Third World markets, the promise to cut subsidies on exports, particularly agricultural exports, has not happened. "Instead, 'market access'—the cutting of quotas and tariffs—allowed agribusinesses in the developed countries to dump subsidized imports on developing nations to gain control of those domestic markets by competing against domestically produced food grown by unsubsidized farmers."[61]

We will explore in depth the history and legacy of the US public sector support for agriculture in Chapter 4; however, at this point it is critical to understand how important agricultural exports are to the US balance of trade. While the United States has to compete in fields such as electronics, it has been able to maintain dominance in agricultural products. Second only to weapons, the number two US export is agriculture. It is the world's largest exporter of maize (corn), wheat, and soya, and only second in rice.[62] The United States is able to maintain this dominance through government subsidization of their grains. However, subsidization results in overproduction that the US domestic market is unable to absorb. More than one in three acres of US crops are exported in bulk or in value-added forms.[63] This level of overproduction makes exportation of these surpluses necessary either by dumping these grains in other agricultural markets or disguised as aid. For instance, during the 1991–1992 drought in Southern Africa, the United States dumped highly subsidized wheat into Zambia at a selling price cheaper than a break-even price for local producers. "Zambian wheat, which should have elicited premium prices because of the drought could not be sold."[64] The Zambian agricultural markets had been forced open by SAPs during the 1980s and had no mechanisms to block this dumping. By 2003, core countries were "spending some U.S. $300 billion a year on farm subsidies, about six times more than on development aid."[65]

Prior to the WTO, the United States had been able to prevent regulation of their agricultural subsidies and dumping by negotiating exceptions to agriculture under GATT. But with advent of the WTO, dumping of subsidized grain by core countries was supposed to come to an end. However, as of 2012 core countries have been able to maintain their subsidies through an elaborate shell game using boxes that allow exceptions for some agricultural policies. The shell game is used to "hide" the subsidies under policies for domestic agriculture that are categorized by the effects they have on production and trade or what is referred to as Overall Trade Distorting Domestic Support (OTDS), which is measured by the total amount of support as calculated by a country's Aggregate Measurement of Support (AMS). There are three boxes: amber, green, and blue. A policy that has substantial impact on trade is placed in the Amber Box, and is automatically added into the AMS calculation and must be frozen and reduced under AoA rules. The game is then to switch subsidies to the green or blue boxes.[66] The United States and EU have made extensive use of the green box, which includes polices that are not supposed to have a major effect on production or trade. Green box programs such as "warehousing food against famine, agricultural research, pest eradication, agricultural training programs, food and agriculture inspection programs, marketing and promotional programs, infrastructure and direct payments not connected to agricultural production, such as income insurance and social safety-net programs and environmental programs"[67] are not subject to reductions. Whereas some of these programs are clearly subsidies, they are considered "decoupled" subsidies because they supposedly do not affect current production and prices. For instance, the US Farm Bill in 2002 used the green box to hide an additional US$189 billion support for its domestic producers in the form of export credits and food aid.[68] Blue box programs are those that link government payments specifically to production-limiting programs and can fall under OTDS limits if they are not decoupled. They are considered decoupled if they provide income support on a fixed number of acres or animals, and with production-limiting features such as land set-asides that are not linked to current production and prices.[69] So, although only green box measures are automatically exempt from inclusion in a country's AMS, blue box measures can be considered "decoupled income support" that is not reflected in the commodities' prices and thus also can be excluded from the AMS. "Perversely, these exclusions mean that incomes of some farmers in countries wealthy enough to do so will be directly paid by governments and will thus be insulated from trade or market fluctuations."[70] This ability to be protected from free market fluctuations is not open to countries in the developing world because of lack of resources and certainly does not fit within the neoliberal mantra of free trade that the WTO was purportedly set up to create.

The Global South Finds Its Voice and Fights Back

The failure of the Uruguay Round to end these obfuscations by core countries was causing increased resentment in the periphery countries and in Seattle, December 1999, the pot boiled over. Seattle was to be the WTO's fourth ministerial-level summit called to resolve the sharp division that had emerged over agriculture, in addition to whether the WTO should expand to additional areas. These new areas in investment, procurement, competition policy, and trade facilitation were first considered when the rich countries tried to launch them in the 1996 Singapore WTO ministerial. Leading up to Seattle, developing countries still refused negotiations on these issues, insisting that negotiations be restricted to resolving the agricultural issues that continued to harm their economies. They were also becoming increasingly distrusting of the secretive processes that were used to make these critical decisions. Historically, most important WTO decisions had been made in closed-door "Green Room" sessions in which only core countries' representatives were allowed, along with a few developing countries that had been co-opted to place a thin veneer of democratic accountability over the negotiations. Seattle was no exception to this and invoked seething discontent among many developing countries at being sidelined from the decision-making process yet again.[71] However, this was not 1994, and the developing countries, buoyed by media attention brought by tens of thousands of activists in the streets outside, were not about to remain silent this time. For more than two years, activists in scores of nations had been protesting the proposed WTO expansion and the failure to end subsidies and dumping. Networks of social movements such as Our World Is Not for Sale Network and civil-society campaign groups had been preparing for this showdown. "Midday on December 3, 1999, blocks of African, Latin American, and Caribbean countries issued public statements, aligning themselves with longtime opponents of WTO expansion like India and Malaysia."[72] The crowd of some 30,000 people outside the convention center— youth, Indigenous groups, international and environmental activists, steelworkers, teamster unions, and Seattle residents—roared back as the megaphones passed the news that the enemy had been named, it was corporate-led globalization behind the mask of the WTO, and it had been stopped. Transnational corporations had finally met grassroots democracy and the people had won.

Industria Changes Tactics and Survives

However, this was only one battle and *industria* was not about to send up a white flag. *Industria* cannot not stop sucking up the world's resources and enclosing the global commons for profit, because if it does not continue to grow, it will die. Surrender is never an option, it must be defeated. So in a cynical attempt to assuage the concerns of developing countries the WTO announced the launching of a new round of talks in 2001 by referring to it as the "development" round. Set in Doha, Qatar in November 2001, immediately following the terrorist attacks of 9/11, this round of trade negotiations was sold as the "development round" designed to place the needs and interests of the poorest members of developing countries at the heart of the talks.[73] Originally, it was to be completed by 2005, but has continued to be plagued by unrest with the ministerial in Cancun 2003, failing due to "US and European unwillingness to be more forthcoming on agriculture."[74] In the summer of 2008, the stalemate was the same, the modalities for agriculture and Nonagricultural Market Access (NAMA), desired by core countries, had failed, putting a cloud of uncertainty over other issues such as rules, services, and TRIPS.[75] Pascal Lamy, the WTO director-general, in an attempt to force through the Doha WTO expansion, went beyond the normal closed-door "Green Room" of around thirty-some members from primarily core countries to restrict final deliberations to seven regions—the United States, EU, Brazil, India, China, Japan, and Australia. This supposed "development" round left out all of Africa to say nothing of the rest of the 146 member countries and concluded with a take-it-or-leave-it deal to the "Green Room" invitees—they left it.[76] In what Wallach contends was a victory for small farmers, workers, civil society, and the developing world as a whole, the primary areas of division remained the same. The core countries tried to maintain their subsidies through the use of green and blue boxes, while demanding lower tariffs in nonagricultural markets in the developing world. Developing countries wanted these subsidies addressed and an effective special safeguard mechanism (SSM) put in place to protect their import markets, farmers, and national food security from excessive import liberalization.[77] Negotiations in services had also been forestalled with a group of countries—Bolivia, Cuba, Venezuela, and Nicaragua—circulating a "proposal to remove health care, education, water, telecommunications, and energy from the WTO, on the basis that these essential public services are human rights which governments have an obligation to provide, and should not be treated as tradable commodities."[78]

Deborah James from the Center for Economic and Policy Research characterizes the breakdown as follows: "Negotiators threw in the towel on their seven fruitless years of trying to expand a particular, corporate-driven set of policies, to which the majority of governments have said 'no,' time and time again." She continues that the collapse went beyond these trade issues to include "issues surrounding the food, climate and financial crisis—as well as the lack of progress on development due to the failure of neoliberal policies to actually promote growth or reduce poverty."[79] There seems little hope that when the 157 members of the WTO meet in Bali in 2013 there will be a break in the impasse over subsidies or other major issues. Lamy warns against expecting too much. "I am neither under any illusion that the factors that have shaped the impasse which we face have changed substantively nor do I harbour any dream about achieving grand designs or comprehensive deals."[80]

The 1993–1994 change in development policy in the United States and core countries from "trade not aid" has failed, not only for developing countries but also in core countries. Social services have also been cut back in core countries of the global North even as factories flee to countries without social regulations and low wages. Natural resources are being ripped from the Earth with pollution spewing into the atmosphere from transporting them from every far-flung corner of the planet. Countries once self-sufficient in food are now forced to import food at global prices, again transported halfway around the globe. The United Nations Food and Agriculture Organization index showed that global food prices rose 12 percent in 2006, 24 percent in 2007, and 50 percent over the first eight months of 2008.[81] The Bank has estimated that rising food prices had pushed 100 million people below the poverty level and have sparked protests and even riots in some parts of the world.[82] "By 2009, the total number of hungry people in the world had topped one billion." The FAO report continues that "by December 2010, the FAO Price Index had topped its 2008 peak."[83] Droughts in the US Midwest in 2012 have already raised food prices for the end of 2012 and into 2013. Unfortunately, under warming conditions, these climate anomalies, like the US heat and drought conditions in 2012, are expected to become norms in the twenty-first century. Battisti and Naylor have shown that there is a better than 90 percent chance that average growing seasons will exceed the heat extremes of the twentieth century, which will mean a loss of between a fifth to a third of world food yield, and severe food crises in areas where three billion impoverished people live.[84] Worse, as temperatures reach 30–31°C, photosynthesis for many crops such as wheat crosses a threshold and essentially shuts down, and this may mean a loss of anywhere between 30 and 82 percent declines in food yields.[85] The Bretton Woods international institutions' neoliberal economic policies have selectively benefited *industria* and its corporate sponsors to the detriment of the people of the Earth and the very ability of the planet Earth to support all those who depend on her. It is time to search for an alternative model of development and planetary consciousness to guide it.

Chapter Four
Capitalism's Endless Pursuit of Profit Destroys the World's Food Supply

> The Western conception and practices of technology are bound up in essentially human-centered materialism: the doctrine that physical well-being and worldly possessions constitute the greatest good and highest value in life. Indigenous conceptions and practices of technology are embedded in a way of living life that is inclusive of spiritual, physical, emotional, and intellectual dimensions emergent in the world or, more accurately, particular places in the world.
> —*Daniel Wildcat (Yuchi member of the Muscogee Nation of Oklahoma)*[1]

Immanuel Wallerstein contends that to "the extent that we each analyze our social prisons, we can liberate ourselves from their constraints to the extent that we can be liberated."[2] Our social prison, as constructed by rhetoric based in neoclassical economics, tells us what our values must be and proscribes a course to fulfill what can only be material values. The course is corporate-led globalization touted by the carnival barkers of *industria*. In this chapter, we will examine this myth sanctioned by neoclassical economics and explicate the ramifications of following this path in harms to both the planet Earth and the rich diversity of cultures enjoyed by the human community. In particular we will focus on the effects of corporate control of the global food supply that is destroying the ability for local communities to decide what foods they wish to grow and consume. We will demonstrate how this modern industrial agriculture is destructive to the natural environment, human and nonhuman health, in addition to the cultures of local communities. Currently, industrial agriculture tells us either we adopt these policies or starve. As with the larger project of corporate-led globalization, we are told that there are no alternatives to this model with its sole goal of increasing economic growth.

Capitalism's Crisis of Overproduction

To begin our discussion, we first turn to Nicholas Faraclas, who contests the accepted wisdom that corporate takeover of the global economy is beneficial to all and is inevitable. Instead he characterizes it as "the most massive transfer of wealth in the history of humanity from the poor to the rich sponsored by the World Bank/International Monetary Fund ... and enforced by the World Trade Organization."[3] His critique of the Bretton Woods international organizations traces their drive to gain access to ever greater resources in pursuit of corporate profit from World War II to the turn of the twenty-first century. Faraclas has named the enemy, it is the corporate-led capitalist model of development supported by its handmaidens, the Bretton Woods institutions. This is *industria* that pursues the imperative of endless accumulation of capital without which it will collapse of its own hollow core. The core is hollow because the global economy erected by *industria* has reached its limits to produce value. Walden Bello contends that it is firmly locked into one of the major "contradictions" of capitalism, "the crisis of overproduction, also known as over accumulation or overcapacity."[4] The current crisis of overproduction was exacerbated by the slowdown of economic growth in the mid-1970s, resulting in stagflation or low growth combined with high inflation. Bello lays out three escape routes taken by those who control the levers of the global economy. To spur growth the first escape was neoliberal reforms pushed by Reaganism and Thatcherism in the global North accompanied by structural adjustment programs in the global South. For Bello the major weakness of this strategy was that, in redistribution of income to the rich, the incomes of the poor and middle classes were gutted. When the wealth did not "trickle down," demand was restricted. There were not enough people with income to buy this overproduction.

To counter this continued overproduction, the next strategy became globalization, or in Bello's term, extensive accumulation. Under globalization, rapid integration of semi-capitalist and subsistence economies became the model of the 1990s. Access to cheap labor and raw materials was once again able to drive profits for a short time as it had done in the colonial period. However, as the twentieth century drew to a close it was clear that this escape route from stagnation only was exacerbating the problem of overproduction because it also added to productive capacity.[5] Financialization, the third escape identified by Bello, proved even more destructive to the global economy. "With investment in industry and agriculture yielding low profits owing to overcapacity, large amounts of surplus funds were circulating in or were being invested and reinvested in the financial sector—that is, the financial sector was turning on itself."[6] While the real economy remained stagnant, the financial economy, which was producing nothing of value, created a "radical rise of prices of an asset far beyond real values."[7] This creation of a bubble was seen very clearly in the US housing market with the subprime implosion in 2008. The demand for housing had been artificially created by those who wished to make a quick profit on speculation that these mortgages could not be paid off. The game in the financial sector became to securitize these bad loans and sell them as quickly as possible to the next person while the seller skimmed off his profit. Although those first sellers in the financial sector made a profit, no value had been created and the demand never really existed. When interest rates on subprime loans, adjustable mortgages, and other housing loans increased, the bubble burst, infecting the whole global economy. In response, banks stopped lending and began buying up rivals. Bello's point is that none of these three escape routes worked because of the problem of over-accumulation. All of them led to increasing levels of overproduction without creating any new value.

States Protect Corporate Profit

This critique of capitalism's pursuit of endless accumulation leading to implosion has been articulated by many. For our purposes, we will focus on two primary scholars, Immanuel Wallerstein and Rosa Luxemburg. Wallerstein in his world-system model takes a longer approach to the problem of diminishing profits. He contends that the capitalist world-economy, evolving since the 1600s, has always been driven by the imperative of endless accumulation of profit, which in turn "generated a need for constant technological change, a constant expansion of frontiers—geographical, psychological, intellectual, scientific."[8] Globalization is only the most recent expansion of frontiers designed to refine the capitalist world-economy's drive for profit. The only difference this time is that profits are getting harder to come by. For Wallerstein, the capitalist model does not gain its profit from efficiency as the neoclassical model argues, but rather from states privileging some corporations with the power to become quasi-monopolies and helping them defray some of the costs of production. Strong states in this system are created by the ability to provide these services to corporations. "There exist no neutral positions for the state in enabling capital accumulation. For capital accumulation is always capital accumulation by particular persons, firms or entities."[9] The awareness that states must pick those who benefit from their embrace shines the spotlight clearly on the intense lobbying process in legislative bodies of democracies. This lobbying money is clearly well spent.

Profits can come from raising the price of goods or cutting the costs of production. Because of competition, combined with a lower level of effective demand due to less money trickling down to potential consumers, raising the price of goods is becoming more difficult. As a result, corporations are turning to reducing the costs of production. In Wallerstein's capitalist world-economy, states play the key role in setting up conditions that allow the quasi-monopolies increased opportunities to choose this strategy to augment profits. Various strategies are designed around the elements of production, which include personnel, taxes, in addition to the often hidden inputs—waste disposal, renewing raw materials, and infrastructure costs.

A primary strategy for reducing costs of personnel under corporate-led globalization is to move production out of states that provide protection for labor, including health and safety measures and the right to organize. Strong states impose no penalty on corporations for this outsourcing of jobs by moving their operations offshore to take advantage of cheaper labor.[10] The main target areas of these "runaway factories" are nonurban areas of the world where people practice subsistence living. Because the workers are not in the wage economy, even the small wages offered initially can be seen as a "significant increase in the overall income of the household of which they are a part, even if the wages are significantly below the worldwide norm of remuneration."[11] However, this market incentive only works to convince some subsistence farmers to abandon their collective ways of life and, even for those few, it has limited effect over time. For Wallerstein, this strategy to pursue cheap labor eventually fails as the workers themselves or the next generation begins to recognize how their wages compare to worldwide norms and start to engage in syndical actions, organizing to press for increased standards of remuneration. As the world continues to de-ruralize there will be fewer places for factories to run.

A second cost of production that states can provide to these quasi-monopolies is tax reduction. However, this too has limits for Wallerstein because of the currently low rate of corporate taxation in the global North. For example, the United States has a statutory corporate rate of 35 percent; however, the effective rate due to deductions, credits, and other mechanisms by which corporations can reduce their taxes, resulted in an effective rate of 13.4 percent of their profits between 2000 and 2005.[12] Because each industry has its own exemptions, the Congressional Budget Office in 2005 found that some corporate tax rates were actually negative.[13] The lower corporate tax rate combined with less progressive tax schedules have already resulted in decreases in social benefits in the global North under neoliberal reforms. Populations in the global North are becoming more resistant to further cuts in education, health, and safety.

The remaining option that states can offer corporations is to facilitate externalizing the costs of three major inputs—renewing raw materials, infrastructure costs, and waste disposal. By externalizing these often hidden costs onto the taxpayer, corporations can maintain control over more surplus value for profit. Externalities are any cost of production that is not included in the price of a good or service. We will be focusing on negative externalities, mostly environmental harms. For instance, when you buy a car, the costs of retrieving the iron ore and bauxite out of the ground are included in the price of steel and aluminum used to make the car. However, the costs to find new sources of these materials will not be included and are often picked up by the taxpayers of the state, which then engages in various mechanisms to support corporate needs for replacement of these materials. Historically, empires were erected to secure constant sources of raw materials. Today, states often turn to the Bretton Woods institutions or treaties; as a last resort the military has been deployed to secure these replacements, all at the taxpayers' expense.

The next hidden cost externalized onto the taxpayer is transportation to get the materials to *industria*'s core. As *industria* has extended its tentacles to all corners of planet Earth to suck up the last remaining raw materials, states have competed with each other to subsidize corporate requirements for economies of scale in transport systems. Stephen Bunker and Paul Ciccantell trace how firms and states have "collaborated to develop transport technologies in order to achieve national trade dominance." This collaboration between the Japanese state and firms resulted in new reinforced steel hulls in which the Japanese could "transport iron ore halfway around the world more cheaply than U.S. Steel could ship its iron ore across the Great Lakes."[14]

The third externality is the harmful pollutants left in the environment, and pumped into the air and water during all stages of production, compounded by the destruction of entire ecosystems. In our example of the car, the carbon dioxide, sulfur, and nitrogen oxides emitted from burning the coal necessary to change these raw materials into steel and aluminum are left in the air for the taxpayer to deal with, as are the other environmental and social harms caused by the retrieval and transport processes.

We are not disputing the state's role in research and development to give a state's corporations a technological edge. The point we wish to highlight is that the profit generated by the state's assistance in externalizing these costs goes only to corporations, not the workers, consumers, or the taxpayers who pick up the bill. We will explore the full costs of some of these externalities throughout this chapter.

This neoliberal strategy to increase corporate profit is often referred to as the "Washington Consensus." This term was coined by economist John Williamson to identify a policy consensus emerging in the 1980s. He conducted a poll of the political elite connected with the Ronald Reagan and George H. W. Bush administrations, including "technocratic Washington of the international financial institutions, the economic agencies of the U.S. government, the Federal Reserve Board, and the think tanks."[15] This became the ideal theoretical myth for the corporate elite because of its sole focus on aggregate economic growth as the only solution to world poverty. Earlier development discourse, which included questions of equity and redistribution, were neatly swept away by free market reforms purported to lift all boats by trickling down benefits from this growth to the poor of the world. Robin Broad and John Cavanagh argue that this exclusion of all but the elites in Williamson's poll was akin to restricting a poll in the Ministry of Magic to only those in power. Harry Potter would not be polled.[16] Development discourse had shifted and corporate-led globalization emerged as the only path to address world poverty. The World Economic Forum at Davos became the mecca for promoting this consensus and the Bretton Woods institutions became the chief enforcers.[17]

Wallerstein concludes that rolling back costs of production to increase corporate profits has reached its limit, as have increasing price and lowering corporate taxes in the United States, and that these strategies will have declining effectiveness in staving off system-wide change brought by system crisis. For example, by the turn of the twenty-first century capitalists were increasingly turning to seek profits in the arena of financial speculation rather than production. However, "such financial manipulations can result in great profits for some players, but it renders the world-economy very volatile and subject to swings of currencies and of employment. It is in fact one of the signs of increasing chaos."[18] The worldwide financial crisis begun in the US housing market is consistent with Wallerstein's prediction in 2004 that the current capitalist world-economy has reached a crisis level that cannot be addressed within the system and further crises will continue to emerge until the root causes are addressed.

Other thinkers indicate that the hegemony of the United States in particular and the West generally, is declining in political and economic power. While social crisis may be followed by a temporary belle époque, all core hegemons eventually decline.[19]

Subsistence Economics Challenges Capitalism's Need for Continual Self-Expansion

To help us clarify the root causes of this crisis even further, we draw on the work of Rosa Luxemburg and her insights into imperialism. Her unique, early contribution was to explicate how capitalist accumulation was dependent on noncapitalist economies. A Polish-born theoretician and revolutionary, her work on the accumulation of capital first published in 1913 was a critique of Karl Marx's theory of the process of social reproduction or more simply, capitalism's need for continual self-expansion. Luxemburg challenged Marx's assumption that the capitalist system is a closed model in which both the mode of production and articles of consumption are self-contained. For her, not only did the original capital accumulation come from noncapitalist colonies, but continued exploitation of noncapitalist economies was essential if profits were to be maintained. In other words, profit could not be produced from capitalists selling to other capitalists. Economic growth could only continue if there is consumption of new commodities by *noncapitalist or subsistence producers*. The process of expansion of the scale of production in which even larger quantities of commodities could be sold was dependent on the exploitation and eventual destruction of subsistence economies.[20] For Luxemburg, the capitalist global system can never be closed. "Capitalism needs non-capitalist social strata as a market for its surplus value, as a source of supply for its means of production, and as a reservoir of labour power for its wage system."[21] As June Nash makes clear, she was not just saying capitalism needed a market for its commodities, but needed a "market for its surplus value." Of course capitalism needed the raw materials and cheap labor from noncapitalist economies, but it also demanded that they purchase their overproduction for continued economic growth. Luxemburg further asserts that subsistence-based societies will not willingly become commodity buyers because they realize that this economic form will lead to the destruction of their social units. As such, she predicted that military force will invariably be necessary to "persuade" subsistence-based economies to buy the capitalist surplus commodities.[22] Silvia Federici amplifies this point: "War is on the global agenda precisely because the new phase of capitalist expansionism requires the destruction of any economic activity not subordinated to the logic of accumulation, and this is necessarily a violent process."[23] Others contend that military force may not be required because the key element of hegemony is that subjects volunteer their obedience because the hegemon appears to be a beacon of the future without any alternatives, and the norms of the hegemon come to define commonsense expectations in everyday life.[24] In other words, the hegemon need not physically force farmers to switch to a pesticide model for high-yield seeds for export, instead of subsistence, if a discourse constructs these seeds to be the only apparent rational option to make ends meet in the system that has formed around them. In this case, "volunteering" for intensive Green Revolution practices replaces traditional practices that eliminate both biodiversity and cultures in the process,[25] while the farmers find themselves in a "pesticide trap" of debt, danger from the poisons, and declining chemical effectiveness that acts more like a treadmill going nowhere than a path out of poverty.[26]

Women and Subsistence Production

German feminists Maria Mies and Veronika Bennholdt-Thomsen look back to Luxemburg to extrapolate how capitalism exploits and eventually destroys subsistence economies today. They define noncapitalist economies as coming from a "subsistence perspective or view from below" in contrast to the perspective "from above" that aims at permanent growth of goods, services, and money. On a most fundamental level a subsistence perspective "enables people to produce and reproduce their own life, to stand on their own feet and to speak in their own voice."[27]

Mies highlights the distinctions between subsistence and commodity production as follows:

> Subsistence production or production of life includes all work that is expended in the creation, re-creation and maintenance of immediate life and which has no other purpose. Subsistence production therefore stands in contrast to commodity and surplus value production. For subsistence production the aim is "life," for commodity production it is "money," which "produces" ever more money, or the accumulation of capital. For this mode of production life is, so to speak, only a coincidental side-effect. It is typical of the capitalist industrial system that it declares everything that it wants to exploit free of charge to be part of nature, a natural resource. To this belongs the housework of women as well as the work of peasants in the Third World, but also the productivity of all of nature.[28]

The exploitation of the "free work" of women, peasants, and nature is how capitalism creates its profit. Today, the cost of natural resources includes only the cost of retrieval, not the time nature took to create them or the cost of replacement. The work of women in reproduction of daily life through cooking, cleaning, and nurturing is never accounted for in neoclassical economic measurements of productivity. All of the work of subsistence farmers is excluded from international measures of income. Yet this "free" labor of nature, women, and peasant farmers is what supports the majority of peoples on this planet, ensures a quality of life, and cushions the harsh economic dislocations of globalization as farmers are forced to stop growing food for their own consumption and produce export crops to earn foreign exchange under neoliberal free market reforms. It is the remaining subsistence producers, most often women, who provide what food there is when cash crops fail to make enough money to buy imported food.

Mies and Bennholdt-Thomsen focus a particular spotlight on the invisible work of women under a capitalist economy. They are invisible because their work in reproducing daily life does not go through the market. For instance, the labor of mending a sock, cooking a meal, or rocking a crying baby is not recognized by the market because no money was exchanged. Because this work does not go through the market, it cannot be measured as growth or development. This invisible work under neoclassical market measures has increased the quality of life for the family but does not show up as part of GDP and therefore does not count. Mies gives a classic example of the harm done to women by making their subsistence labor in the home invisible. In her study of poor rural women in South India, she documents their exploitation by lace makers in Europe, Australia, and the United States. These women earned only a small fraction of their worth for crocheting lace in their homes. While the lace exporters became millionaires, they justified paying the women a small pittance of their worth based on the logic that "these women were 'housewives' who were anyhow sitting idly in their houses. By making lace they could use their time *productively*."[29] One can only imagine what critical life-sustaining household labor had to be foregone to engage in this "productive" labor.

As long as measures of development are limited to those based on neoclassical economic growth, subsistence production will continue to remain invisible resulting in untold harm to subsistence communities and the planet Earth. The surest way to ensure the destruction of subsistence economies, whether existing on their own or in combination with market economies, is to accept the myth that economic growth is the only solution to world poverty and to impose some linear, Western view of progress. As profits continue to be harder to come by, there will be increased destruction of these valuable ways of life including the increased use of military power to exploit even the most isolated corners of the world in *industria*'s fevered frenzy to squeeze all profit from life on planet Earth.

But this is not the only alternative path for addressing world poverty. Contrary to the Washington Consensus that there are no alternatives to corporate-led globalization, other voices have emerged to challenge this myth. For example, More and Better Network, a transnational network for support of food, agriculture, and rural development to eradicate hunger and poverty, and La Via Campesina, supporting peasant and Indigenous peoples, have emerged. These are peasant movements mobilizing to valorize the work of subsistence as a crucial practice with deep knowledge and experience that is of value to peasants and to the rest of the world.[30] Peasants and Indigenous peoples of the world are and have been living an alternative to the Washington Consensus, and they are and have been willing to fight for it. These Harry Potters of the world who were supposed to be the beneficiaries of this myth are not waiting to be polled and are organizing around their slogan, "another world is possible." This movement of militants has been engaging in collective resistance as well as rebellion to protest corporate takeover of their economies. These alternative voices on the political left have often been identified with the World Social Forum (WSF).

Despite a dearth of coverage in mainstream media, the WSF has become a site of antiglobalization, anticolonialism, and antipatriarchal resistance. The

WSF began in Brazil with the first three meetings being held in Porto Alegre starting in 2001. Originally, with annual and now biennial meetings around the world, these peoples gather in order to create a space for open, critical, and reflexive responses to modernity's crises through more inclusive politics. The WSF brings together activists who oppose neoliberalism, including subaltern peoples of the world. Subaltern populations are "populations who are practically excluded from modern politics, despite rhetorics of universal inclusion. These include slum dwellers, tribal and Indigenous peoples, Afro-descendents, *dalits* [the so-called 'untouchable' caste of India], subsistence producers, among other poor people's movements."[31] The WSF has specifically been organized in a way that purposefully keeps ideas decentralized, so that no one idea from the WSF emerges to dominate others in a process described as "open space." It is organized, in other words, to maintain the smooth, undominated spaces described by Deleuze and Guattari[32] and Hipwell,[33] as we have discussed in Chapter 1. The WSF is an umbrella movement of movements that are opposed to the economism of Davos and the Bretton Woods system and to the reckless and anti-democratic accumulation that puts the world at large at risk. While the WSF experiences its own contradictions as it attempts to practice global justice, it valorizes the local and place-based efforts to dislocate the universal demands of industrial control with the basic claim that "another world is possible."[34] It is this other world that the neoliberal world capitalist system meeting in Davos has attempted to foreclose. However, even at the WSF, ecological concerns had been marginal until the 2004 meeting in Mumbai, where a large portion of the 80,000 conference attendees were women, *dalits,* and *adivasis* (Indigenous peoples in India). After more than a decade of studying the WSF, Janet Conway writes that the Mumbai WSF meeting

> foregrounded issues central to the survival of tribal peoples: their subsistence rights to lands, rivers, forests and water against the destruction wrought by megadevelopment projects, resource extraction, privatisation and corporate control of nature. These movements are rural, communitarian, oriented to subsistence livelihoods and embody the links between bio- and cultural diversity. Their survival struggles forced ecological questions to the centre of the WSF's agenda which, before and up to the 2009 WSF in Belém, had been relatively marginal. Their presence also posed deep challenges to the modernisation, urbanisation and development discourses that continue to underpin the utopias of much of the "anti-globalisation" movement.[35]

In 2009, the WSF, held in an Amazonian town, Belém, Brazil, brought a large contingent of Amazonian Indigenous peoples to the meeting. From that meeting, Indigenous leaders began organizing a thematic focus on "the crisis of civilization" to highlight the world Indigenous movements' discussion about "the struggle for plurinational states, climate justice, and *buen vivir* (living well, not better) as a critique of consumer society."[36] Some prominent thinkers of the WSF, such as Boaventura de Sousa Santos, believe that the neoliberal world order, and even Western modernity itself, is entering a final crisis, and therefore a period of paradigmatic transition that will alter the very contours of the future. De Sousa Santos believes that the WSF may be the source of new social paradigms, because it has fostered indiscriminate diversity: "the WSF represents the maximum possible consciousness of our time."[37]

Wallerstein asserts that we must make a choice between the values promoted by Davos or Porto Allegre. In this chapter and the next three, we will present some of the voices of the World Social Forum, Indigenous communities, and subsistence peasants who challenge the Washington Consensus model of development in order to allow the reader to understand the bars of their social prison constructed by *industria*. Another world is possible if the young decide they want it.

Subsistence Communities Challenge Neoclassical Economic Conceptions of Individual Profit

A common theme that runs through many of these voices is a defense of their subsistence economies that are being destroyed by corporate-led globalization. James C. Scott's now classic work, *The Moral Economy of the Peasant*, shows that a singular focus on profit maximizing is not compatible with peasant survival strategies, and as this development approach was thrust uninvited upon farmers in Southeast Asia, they revolted to protect their alternative to living well.[38] Eric Hobsbawm has predicted these imposing conditions may mean the end to the peasantry,[39] but it is more likely that the neoliberal world order is recomposing class struggles among the peasants who are learning to fight back.[40] The key to understanding subsistence economies is the vital role played by the peasant farmer. These cultivators can belong to economic units ranging from Indigenous groups to small farmers in developed countries who have rejected the industrial model of farming. Mies and Bennholdt-Thomsen define peasants as "all those who have direct access to the land and do not receive agricultural wages."[41] Unlike the growth economies, this way of life has a different worldview, which "recognizes the finite basis of economic activity in land, water, forests, plants, and the need to operate with corresponding care and restraint."[42] This worldview is consistent with the idea of an ecological community in which the focus is on the interconnections amongst the members of the community rather than the isolated parts to be exploited. This awareness of the limits of nature often extends to a social behavior in which all "members of society feel an obligation to conduct their economic affairs in such a way that others are also able to survive, drawing assurance from the knowledge that their own basis of existence will always be safe."[43]

Stephen Gudeman developed an economic anthropology that helps to elucidate the different values pursued by what he terms a community economy versus a market economy.[44] Based on his ethnographic fieldwork in small rural villages from Colombia, Panama, and Guatemala, he enlarges on his early neoclassical economic training to theorize a community economy that he argues operates dialectically with the impersonal, abstract market economy with its focus on accumulation. He contends that while neoclassical economics does not acknowledge the community economy, it nevertheless is dependent upon it for profit. He bases the concept of profit on innovation that he asserts is always derived within a cultural or communal context. The community economy, typically nurtured in a subsistence way of life, is composed of both a base and social relationships based on trust that sustain the base. The base is formed by a community's *"shared interests,* which include lasting resources (such as land and water, produced things), and ideational constructs such as knowledge, technology, laws, practices, skills, and customs."[45] These construct a commons developed through social relations that are maintained as an end themselves. The commons is not a common-pool resource to which access can be controlled for only economic reasons. Rather it is for Gudeman a social sphere of value that has material and knowledge components that are shared by the community. Because the central process is making and sustaining a commons, subsistence economies exercise great thrift in caring for the base. This value of conserving the base extends to a household's use of resources. "A household exercises parsimony in its consumption of foods and use of goods; people try to prolong the life of items in use and to expend as little as possible."[46] The community economy dialectically interacts with the market economy, which allows for impersonally traded goods and services and accumulation of wealth or profit. For Gudeman, profit is a result of the innovations within the community economy, which is ignored by neoclassical economics.

To clarify his theory of profit, Gudeman presents an overview of the evolution of the sources of profit from the mercantilists to the neoclassical economists. Only the neoclassical assert that profit comes from within the market system generated by firms finding increased sources of efficiency. For the mercantilists, the source was gold, for the French physiocrats it was the reproductive parts of nature set in motion by divine power, and, originally, Ricardo said it was the return on the inherent qualities and powers of the land.[47] He later developed a human labor theory of value, eventually to be refined by Marx. For these theorists the market domain did not create value, so "fresh value had to be rendered as an external gift to the existing system of value."[48]

Modern neoclassical theory presents a closed model in which firms are the key actors who use instrumental rationality to select the most efficient means of production. In this model it is innovation by isolated entrepreneurs that creates profit. Gudeman contests this individualistic notion of innovation arguing that "innovations are always set within a community, because the learning and accomplishment are dependent on a social context."[49] He points out that even "Silicon Valley is a net-work, a community of corporations with porous borders—information, people, and teams cross between the profit centers: and the creation of a larger, communal base enhances the market performance of each participant."[50] Yes, there are individuals who excel at drawing on this base and are able to accumulate profit, but the source of value comes from knowledge and innovation developed and nurtured in the base. He gives several examples of how the community base provides the source of innovation in small rural communities. One example is of weavers in Guatemala. This weaving is done by women in the home who draw on the legacy of weaving methods and techniques for which their towns are famous. As the weaving has evolved various weavers produce new designs, which are then copied by the other weavers who in turn add their innovations. The innovations in the designs are thus contributing to the total knowledge base of the entire community of weavers. However, when one house was co-opted by a corporation to expand its operation for sales to the United States, their innovations were no longer shared by the community. The generations of shared knowledge created in the base now became solely the profit of the global corporation, which had created "an enclave industry in which they used external capital and low-cost domestic labor to produce a commodity for export."[51] No one in the community was allowed to see the weavings produced in the enclave. As such, the base was not maintained by this household's innovations. Corporate-led globalization not only rapes the earth, exploits poor workers, and despoils ecosystems, it also destroys the way of life in noncapitalist or mixed economies. Like a vampire, it swoops into subsistence communities, sucks out the innovations created by the community, leaving nothing but waste behind.

Indigenous Alternatives

It should be clear by now that not all remain silent in face of this devastation. Some of the most penetrating voices of resistance are emanating from Indigenous peoples. Nash concludes from more than forty years of ethnographic studies of Mayan Indigenous communities in Latin America that "indigenous peoples will become the chief protagonists of change."[52] Nash bases this claim on the fact that they present an alternative paradigm to capitalism's focus on competition and accumulation. She writes, "Their logic of cooperative effort among members of the same community, and between humans and the cosmological and natural forces, mediated through civil and religious *cargos,* promotes reciprocal exchanges that run counter to private individualistic exploitation of the environment."[53] She does not ignore the conflict within these communities, especially as it revolves around the equality of women, but she maintains that their resistance, presented in moral terms and the right to survive in their traditional collective ways of life, offers the clearest challenge to the instrumental rational logic of free market globalization.[54]

Her case study of the Zapatista movement in the state of Chiapas, Mexico, is a clear illustration of this point. Indigenous peoples of the Americas have been engaged in 500 years of struggle to maintain their way of life in the face of extreme violence. In recent years their resistance has centered on the penetration of corporate-led globalization into their communal economies. In the period from 1988 to 1993, subsistence-based Mayan farmers were peacefully protesting to protect their communal way of life in the Indigenous highlands of Chiapas in southern Mexico. Due to population pressures and actions by the Mexican government to deny their autonomy, many had moved into the Lacandon rain forests in order to have land on which they could maintain their identity and values. Their peaceful resistance erupted into violence on January 1, 1994, when the Zapatista Army of National Liberation (EZLN) led an armed force of some 2,000 men and women from the Lacandon forest and seized several town halls surrounding the entrance to the forest. It was "the culmination of months of peaceful protests and years of gathering resentment to the government's denial of human services and political rights to the settlers in the Lacandon rain forest."[55] After almost two weeks of combat with 12,000 federal troops, then-president of Mexico, Salinas, called for a cease-fire and promised to negotiate with the leaders of the EZLN. The government never did negotiate in good faith and proceeded to invade the Lacandon forest on February 9, 1995. However, this Mayan movement had gained an international audience for its right to survive as a pluriethnic community. At one point in the failed negotiations it even held a worldwide plebiscite on the relations between Indigenous peoples and the Mexican State. One and a half million people worldwide responded to the plebiscite supporting the Zapatistas' rights to autonomy.[56] Their raised profile strengthened their position and helped to force the Mexican government back to the bargaining table, but in the failed negotiations with the Mexican state, the EZLN simply established self-government, or the *comunidades autónomas en rebeldía* of Chiapas that continue as of this writing.[57] Interestingly, the EZLN "made clear that they did not seek to seize state power but to construct a new world within many other worlds in order to pursue *buen vivir*."[58] The fact that *buen vivir*, or a good life in harmony with the Earth and other people, is a consistent demand from resistance movements, tells us that there is an alternative that is, at least in their measure, incompatible with the globalist status quo.

According to Nash, the Zapatista cry was for autonomy, not just in a narrow political sense, but the moral right to their own identity that could only be possible if they had the material basis needed for their way of life. That included communally owned land, rights to have regional institutions to resolve agrarian conflicts, a percentage of Mexico's profits from the oil under their land, and the rivers running through it that produced 52 percent of the electricity for the Mexican nation.[59] This Zapatista revolution has often been referred to as the first postmodern revolution because of its rejection of cultural universals. In seeking recognition of their pluriethnic community they say, "We are all Mexican, but each lives and feels his/her Mexican-ness differently."[60]

Another strong voice presenting alternatives to the neoliberal approach is La Via Campesina, noted above, literally translated from Spanish as the peasant's path. La Via Campesina is an international movement of peasants, small and medium-sized producers, landless, rural women and youth, Indigenous peoples, and agricultural workers. Their 148 member organizations come from sixty-nine countries located in Asia, Africa, Europe, and the Americas. Their primary objective is to develop solidarity and unity among members in pursuit of preservation of land, water, seeds, and other natural resources and support for sustainable agriculture.[61]

These voices had been virtually drowned out by the din of the Washington Consensus until the protests in Seattle in 1999 against the WTO. Together with other activists from around the world they slowed down the neoliberal agenda. While some were temporarily lulled into complacence by the promises of the Doha Round of WTO talks billed as addressing their concerns, most remained steadfast and continued to resist. Their concerns once again were amplified when the world faced the international food crisis in 2008.

Endless Drive for Corporate Profit Destroys the Global Food Supply

The global food crisis of 2008 was not so much a shortage of food as it was, according to Peter Rosset of La Via Campesina, a global crisis of rising food prices, which led to food riots in parts of Asia, Africa, and the Americas.[62] Allowing people to face starvation in spite of adequate food supplies that are held back for financial reasons was well documented by economist Amartya Sen in investigating the 1943 Bengal famine, in which over 3 million people died of hunger.[63] There was plenty of food locked up in the granaries, but it was being hoarded to raise the price. "Those who died in the street died because they simply weren't able to pay enough for the food."[64] In 2007, there was also financial speculation driving up the price of food. As with financial speculation in the housing industry, the financial sector was speculating in agricultural commodities leading to a peak of food prices in June 2008. "Attracted by high price volatility in markets, since they take their profits on both price rises and drops, they bet like gamblers in a casino."[65] Rosset concludes that they were gambling with ordinary people's food by injecting an additional $70 billion into food commodities, inflating a price bubble that "pushed the cost of basic foodstuffs beyond the reach of the poor in country after country."[66] This spike in food prices was especially felt in developing countries that had become net food importers due to the policies of the IMF and World Bank as discussed in Chapters 2 and 3. By 2008, roughly 70 percent of developing countries were net food importers. Of an estimated 845 million hungry people in the world, 80 percent in 2008 were actually small farmers and subsistence peasants.[67]

While the stated purpose of corporate globalization is to increase economic growth thereby raising per capita income, the opposite happened in most Third World countries. A study by the Washington, DC–based Center for Economic Policy Research (CEPR) demonstrated that during the initial two decades of globalization (1980–2000) 77 percent of countries saw per capita rate of growth *fall* by at least 5 percent compared to the previous twenty years (1960–1980) "when many poor countries were focusing on their own productive capacity and meeting local needs."[68] A follow-up study by CEPR "found that progress in reducing infant mortality, reducing child mortality, increasing literacy and increasing access to education has all slowed during the period of corporate globalization, especially in developing countries."[69] The policies of forcing Third World countries to give up growing food for local consumption in order to grow cash crops for exports left them with no food and not enough money to buy imported food when financial speculation spiked the prices of imported food commodities. Compounding this crisis was the inability of some governments to turn to their public sector grain reserves because the IMF and World Bank had forced them to sell off the reserves as part of the neoliberal reforms.[70]

Faced with this crisis, it would seem obvious to all that the industrial agricultural policies of corporate-led globalization were not working and alternatives should be sought. However, common sense did not rule the day; agribusiness, which depends upon new markets in the global South countries for its overproduction, drove the debate once again. Since the beginning negotiations of the Uruguay Round for the creation of the WTO, the US position has been that US agriculture should feed the world. In 1986, at the start of the Uruguay Round, US Agriculture Secretary John Block, in a moment of candor, stated the US position clearly: "The idea that developing countries should feed themselves is an anachronism from a bygone era. They could better ensure their food security by relying on U.S. agricultural products, which are available, in most cases at lower cost."[71] Block was accurate in his prediction that agricultural products would be cheaper from the global North, but only because these countries never eliminated the subsidies given to their producers as was promised in the Uruguay Round.

Industrial Agriculture Brings a "New" Green Revolution to Africa

The neoliberal solution to the global spike in food prices is more industrial agriculture for all countries. Though it is conceded that developing countries, which are now net food importers, were hit hardest, net food exporting countries made huge profits. Neoliberal wisdom dictates that the solution then must be for developing countries to increase industrial agricultural practices[72] in order to take advantage of the higher agricultural prices.[73] Today, in Africa the solution is said to be a new Green Revolution, with enormous funding from the Bill and Melinda Gates Foundation (discussed below), this time with genetically modified organisms (GMOs) added to the mix. Before we accept this prescription we need to assess the long-term effects of the first Green Revolution in Asia and Latin America, which was the most organized assault against peasant agriculture in history.

Agribusiness and its supporters have refused to acknowledge the many environmental and social costs incurred in the first Green Revolution. Their failure to acknowledge these externalities is, as argued by Wallerstein, tied to their dependence on the ability to externalize these costs onto the public in order to realize a profit. The first Green Revolution began in the 1950s with the development in Mexico by Norman Borloug of a dwarf variety of wheat and rice that could increase yield through application of large inputs of chemical fertilizers and water. Unlike indigenous varieties of wheat and rice that would grow too tall and "lodge" or fall over with increased fertilizer, these new High Yield Varieties (HYVs) could increase yield without having to worry about lodging or even the health of the soil, at least for a few years. With the support of the Rockefeller Foundation, and later the Ford Foundation, the American government and international agencies (especially the World Bank, which was more than happy to build more dams), this American model of science-based industrial agriculture was implemented as a purported solution to hunger and poverty. However, "increased concentration of power and control over the food system by transnational corporations," and profit were the true goals of the Green Revolution.[74] Early implementation of this technology in the 1960s occurred in India, particularly the state of Punjab, which had 30 percent of India's agricultural land.[75]

The basic model was to select larger landholders and initially subsidize their access to the expensive miracle seeds and chemical fertilizers—nitrates, phosphorus, and potash (NPK). Traditionally, subsistence farmers share their seeds and therefore incur no cost for them. Another major new cost of this model was the increased amounts of water necessary. HYVs of wheat require at least three times the amount of water as traditional seeds, which required a huge investment by the Indian government in water development.[76] In the first few seasons, grain production increased, but as the soil degraded, farmers needed ever increasing amounts of NPK just to maintain the yield, in addition to ever increasing amounts of herbicides and pesticides that result from monocultures. By the mid-1980s, Indian scientists were calling for the end of monoculture in Punjab because "in terms of resources and energy, the productivity actually declined."[77]

If you do not include the costs of the externalities generated by the Green Revolution in India, it would be fair to conclude that the production of internationally traded grain did increase, but at a very high cost to the environment, health of the people, and their culture.[78] And, it is clear that the Green Revolution is responsible, in part, for overall decreases in extreme malnutrition internationally, except in Africa; however, these benefits are admitted even by Green Revolution enthusiasts to be "uneven" because monoculture cereals make vegetables (key sources of vital micronutrients) more scarce and expensive.[79] Thus, in some places extreme malnutrition decreased but minor and intermediate malnutrition *increased* because, "although overall calorie consumption increased, dietary diversity decreased for many poor people, and micronutrient malnutrition persisted. In some cases, traditional crops that were important sources of critical micronutrients (such as iron, vitamin A, and zinc) were displaced in favor of the higher-value staple crops."[80]

Some of the environmental costs were the harms done to the soil by the high use of NPKs that destroyed the micronutrients in the soil, which was further degraded by salinization from irrigation. The soil also no longer could benefit from the taller stalks of wheat that had not only provided fodder for animals but additionally returned organic matter to the soil. The monoculture model and loss of seed diversity generated new insect pests and diseases that required large infusions of pesticides and herbicides.[81] Inputs of chemicals only work for a limited time, so farmers had to steadily increase the amounts of NPK, pesticides, and herbicides to see an effect. However, subsidies for these inputs were no longer provided, and farmers began to fall into debt from which they could not recover. This has led to a rash of suicides as farmers began to drink their pesticides. Shiva writes of the epidemic of suicide among farmers in Punjab where "vast stretches of land have become waterlogged desert" and "trees have stopped bearing fruit because of heavy use of pesticides that has killed the pollinators—bees and butterflies."[82] "According to the Indian government, between 1993 and 2003, more than 100,000 bankrupt Indian farmers committed suicide." More than 2,000 of these were in Punjab alone, mostly smaller landholders who faced not only indebtedness but "massive social inequality" brought by the Green Revolution.[83] "In Punjab, the epicenter of the country's hi-tech 'Green Revolution,' the United Nations scandalized the government when it announced that, in 1995–96, over a third of farmers faced 'ruin and a crisis of existence.... This phenomenon started during the second half of the 1980s and gathered momentum during the

1990s.'"[84] Not all farmers kill themselves to escape debt; some sell their kidneys. In one village, they have even set up a "Kidney Sale Center." As one farmer said, "Kidneys are all we have left to sell."[85]

Two of the most damaging long-term environmental impacts are the dramatic depletion and poisoning of water in India.[86] To obtain the required water for the HYVs, the "Indian government subsidized the digging of tens of thousands of tube wells to pump irrigation water to the surface. Over the last decades, tube wells have pumped many water tables dry."[87] In Punjab "nearly 80% of groundwater is now overexploited or critical."[88] What water remains is heavily poisoned from the intensive use of pesticides. In 2007, a highly respected research institute completed a two-year study of villages along Punjab's major rivulets. Some of the results include toxic levels of lead, cadmium, mercury, arsenic in many vegetables grown in the area; pesticides beyond permissible limits in vegetables, fodder, human and bovine milk, as well as blood samples; 65 percent of blood samples showed DNA mutation; in addition to a sharp increase in cancer, neurological disorders, liver and kidney diseases, congenital defects, and miscarriages.[89] The author laments that for "Punjab's prosperous farming households and lush green fields, the famed Green Revolution is beginning to turn bilious from within."[90] Cancer rates have become epidemic in Punjab and other parts of India that are using the industrial model of agriculture. In the southeastern Malwa region, "better known as the cancer belt," the farmers are spraying their fields with pesticides far above the recommended dose in order to counter increasingly resistant pests. One farmer admitted that while the "recommended dose is about five sprays per season, we sometimes spray our fields 25 to 30 times."[91] Not only is this poisoning the land and water, it is bankrupting the farmers.

Other externalities not included in the measures of success for the first Green Revolution include the disruption of social ties and ways of life caused by economic restructuring. By 1984, Punjab was being rocked by conflict that Shiva contends was the result of the centralization of political controls necessary to subsidize inputs, which eroded cultural norms and led to an "epidemic of social diseases like alcoholism, smoking, drug-addiction, the spread of pornographic films and literature and violence against women."[92]

If these horrendous costs are internalized into the "yield per acre of internationally traded grain" abstract neoclassical economic measurement of the productive capacity of the Green Revolution, India's people and environment have and are continuing to pay a heavy cost. Consistent with Wallerstein's analysis, it is only by the Indian government facilitating the externalization of these costs onto the people of India that agribusiness can make a profit. Agribusiness will not give up this profit of its own accord.

Subsistence Agricultural Alternatives

To more clearly explicate what choices were involved in pursing this industrial model of agriculture it is necessary to compare it to peasant and/or subsistence agriculture. To begin our discussion we need to understand the difference between yield and output. As Shiva clarifies, yield usually refers to production per unit of a single crop. So yes, measures of yield per acre of wheat did go up during the Green Revolution over the yield of wheat in peasant agriculture. However, if you measure productivity of land by output, which refers to the total production of diverse crops and products from the same land, instead of just yield, peasant agriculture is much more productive.[93] Peasant agriculture most often uses intercropping, or planting more than one crop on the land at the same time. For instance, Mayan peasants in Chiapas only "produce 2 tons of corn per acre, but the overall output is 20 tons per acre when the diversity of their beans and squashes, their vegetables and fruit trees are taken into account."[94] Virtually every objective measure of the productivity of small versus large-scale farming concludes that small farms are more productive. "The USDA, the World Bank, and many other experts have run the numbers and determined that whether the calculation is done on a gross or net basis—before or after the costs of production are subtracted from the value of goods produced—the per-acre productivity of small farms is consistently greater than for larger ones."[95]

These additional foods grown on small peasant farms also insure a more balanced diet and replenish the soil. For instance, pulses (beans) in a traditional diet that is largely vegetarian provide the additional amino acids necessary to make protein when combined with wheat. Pulses also replenish the nitrogen withdrawn from the soil by many crops including wheat and corn.

The diversity of seeds produced in peasant agriculture largely obviates the need to use highly toxic pesticides or fungicides. When you are only planting a monoculture of one or limited varieties of a crop, you are prone to destruction of an entire crop as in the Great Irish Potato Famine (1845–1849). Relying on a small variety of potatoes caused genetic vulnerability. The Irish planted only a small variety of potatoes, which were all vulnerable to the same disease that turned them to mush in the fields within twenty-four hours.[96] Intercropping and rotating crops also keeps the soil healthy, reducing the need for chemical fertilizers and pesticides.

Finally, peasant agriculture does consider yield, but not exclusively. It also grows food for local conditions, taste, and nutrition. Industrial agriculture does not even consider taste, relying on additives such as high fructose corn syrup and salt to replace taste. As a consumer, which food would you rather eat?

The Global Food Supply

That brings our story of industrial agriculture as practiced by corporate monopolies back to not only the limitations forced on farmers but also the controlling of the food supply available to the consumer worldwide. The consumer is not allowed to select from a full spectrum of food choices. The varieties of bread and cereals or quality of beef, pork, or chicken available in supermarkets is totally controlled by transnational agribusiness giants. For example, supermarkets in the global North have only a few varieties of apples that are selected because the skins won't tear as they are knocked about during transportation around the globe. They also are easily waxed so they can look pretty on the shelves and, of course, they "respond well to pesticides and industrial production."[97]

For internationally traded grains these oligopolies select which grains will be grown, using only their seeds and inputs. The farmer cannot make the choice to grow other grains or use their own seeds, primarily because they do not have the ability to transport or process the grains. The global food system requires economies of scale with access to capital to transport these grains great distances to large processing centers. As Patel contends, there are "no mom-and-pop international food distribution companies. The small fish have been devoured by the leviathans of distribution and supply."[98] In developing countries the governments had often facilitated these functions of processing and transporting food from the farm. However, these roles for the governments became some of the first targets of neoliberal reforms in the 1980s and 1990s. Today, farmers have little choice but to do what the leviathans want.

This control of the global food supply from the farm gate to the consumer creates what Patel calls the rot at the core of the food system.[99] Rather than a free market in which millions of farmers choose which crops to grow and how to grow them, driven by the demands of millions of customers, the leviathans of the global food system have created an hourglass shape in which the farmers and consumers at either end can only grow or consume what the monopolies want. And what they want farmers to grow and consumers to purchase is driven by their profit motive. This market concentration has led to higher consumer prices and lower prices for the farmer. In testimony before the US Senate Agricultural Committee, C. Robert Taylor concluded in 1999 that since "1984, the real price of a market basket of food has increased by 2.8 percent, while the farm value of that food has fallen by 35.7%."[100] Commodity prices are lower than they were in the 1970s.[101] The increased cost that consumers pay in the supermarket goes to the corporate monopolies that pick up the food at the farm gate, not the farmer. Since 1992, the average US consumer pays 10 percent of their income for food, but only 2 cents of that dime go to the farmer, if the food is minimally processed. Eight cents go for processing, shipping, packaging, advertising, marketing, retailing, or what the USDA calls "food marketing services."[102] The more a food is processed, the smaller the percentage of the food price that goes to the farmer. For instance, a one-pound loaf of bread retailing at 88 cents means a farmer gets 4 cents, and a one-pound package of potato chips at $3.36 brought the farmer 26 cents.[103] There is a clear profit advantage for these monopolies in highly processed food, in spite of the loss of nutritional value.

The lower price for commodities has worked to drive small family farmers off the land to be replaced by giant industrial operations that plant monocultures requiring all of the inputs we saw in the Green Revolution. Pyle in *Raising Less Corn, More Hell* traces this shift in US agriculture back to government subsidies first put in place during the Great Depression. He holds that these subsides have forced the commodity market into overproduction, which has in turn led to the drop in commodity prices in the United States and increased the pressure to create markets in developing countries for the surplus. In a perversion of the original purpose of the subsidies to save family farms in tough times, government subsidies actually favor the largest producers by basing the current year's subsidy on the amount awarded in the previous year. The larger the subsidy was last year, the larger the subsidy will be for the current year. "Of the $131 billion in subsidy money paid from 1995 to 2003, 23 percent—$30.5 billion—went to a mere 1 percent of the recipients—30,500 of them." Pyle continues, "The top 10 percent of recipients raked in 72 percent of the subsidies—$94.5 billion."[104] Even worse for family farms, these subsidies are going to Fortune 500 corporations such as John Hancock Life Insurance Company, Chevron Oil, and Caterpillar, in addition to wealthy individuals such as banking heir David Rockefeller and media magnet Ted Turner.[105] Small, family farmers do not have the means to expand their operations and continue to fall further behind.

These subsidies also mean that the surplus of internationally traded grain such as corn, wheat, and rice are often dumped into developing country markets below the cost it takes to produce them or given as aid, which can also distort local commodity markets. For instance, in 1998 the "real cost of producing a bushel of wheat in the United States, from seed to export terminal, was not quite $5, but it was being sold for export at $3.50."[106] A bushel of corn costs roughly $3 for the farmer to grow, but "only brings about $2 on what today passes for the open market."[107] Rich countries such as the United States, the European Union, Canada, and Australia provide "some $350 billion a year in subsidies to their farmers. That is seven times what those same nations spend on aid to poorer nations."[108] The US taxpayers are picking up the tab for putting Third World farmers out of business. Farmers in Mexico are driven off their land and Main Street in the United States has lost a tax base that used to be provided by many small family farmers patronizing local businesses.

Even in face of this overproduction, agribusiness is still able to control the discourse with the myth that if the entire world does not turn to industrial agriculture, the poor of the world will starve. The latest iterations of that myth have been used to foist a new Green Revolution in Africa by a joint venture of the Bill and Melinda Gates and Rockefeller foundations called an Alliance for a Green Revolution in Africa (AGRA) launched in 2006. Africa is known for endemic drought, often combined with famine.[109] According to the myth that industrial agriculture with its biotechnology is the only solution, a full frontal assault is being waged against traditional agriculture continent-wide in Africa. Yet, it is this traditional agriculture that has been largely successful in keeping the predicted "millions" from dying with each occurrence of drought. As Mushita and Thompson point out, the predicted "millions" don't starve, in part due to emergency food aid, but more importantly, they are saved by the biodiversity of food sources including many of the 2,000 drought-tolerant indigenous food crops saved in rural areas in addition to some foods designated specifically for famines. "Traditional ecological knowledge designates some highly drought resistant plants to be eaten only in time of dire need. Botswana alone has 250 plants that are used specifically as famine food."[110] Additionally, communities band together to ensure that all members have enough food to survive the crisis. As Gudeman argues, when community economies are recognized, members are more likely to share the benefits of the base with all other members.

It is only through heroic efforts that these indigenous crops have been preserved. Colonial governments first made Africans switch from their "coarse" grains to European "comfort foods" that could also be used as cash crops for export. A second assault on African indigenous crops came from international financial institutions (IFIs) under the structural adjustment programs (SAPs), which conditioned loans on African farmers switching to export crops like coffee, cotton, or cut flowers. In times of food shortages the money derived from these exports has been insufficient to buy imported food that may or may not exist as happened in the food crisis of 2008. Additional conditions of these SAPs demanded that African governments stop support for agriculture under the neoliberal market reforms. In Zimbabwe the 1991 Economic Structural Adjustment Program (ESAP) imposed by the IMF required complete withdrawal of government from guaranteeing food security to their people.[111] It also forced the governments to impose user fees for formerly public services of primary education and rural health centers, during the "worst drought of the century," which hit Southern Africa during the 1991–1992 crop season.[112] Devastating to Africa's small farmers was the requirement that there could be no government regulation of agriculture in "any developing country that needed to borrow funds from the IMF and World Bank."[113] A particular target were the government marketing boards that insured a floor price for grains and facilitated transport of grain to processing centers for small rural farmers who lacked the means to get their crops to market.

The Bill and Melinda Gates Foundation and the connected group, AGRA, fail to recognize any of these structural causes of agricultural decline in Africa. In contrast to thousands of years of African subsistence history that has included the sharing of seeds and knowledge, AGRA focuses on privatizing resources that are part of the global common heritage. Thompson writes, "Removing seed from the public sector and privatizing it are the coercive innovations that AGRA finances. But that is not all. Without genetic diversity, all the privatisation in the world by a few corporations cannot provide adequate food supplies, not for Africa nor for the globe. *The first agenda of AGRA is to facilitate corporate access to the genetic wealth on the African continent, while transforming a long history of reciprocal sharing among farmers into market transactions.*"[114] Further, subsidies from core countries, conditionalities attached to SAPs that require governments to reduce support for agriculture, including agricultural research, and reliance on imports are not addressed in their plans. In fact, the Gates Foundation promotes cash crops "as the pathway out of poverty."[115] Specialization in a few commodities for export increases a country's exposure to volatile commodity prices for both their exports and imports necessary to replace the food production given up to grow cash crops for export. African countries are already vulnerable with nearly a quarter of them dependent on "a single commodity for 50 percent or more of their export income and more than 20 countries rely on two or three commodities for at least half of their export earnings."[116] The sheer insanity of relying on the volatile international commodity market for food imports was demonstrated in 2007 when the price of corn rose by 31 percent, soybeans by 87 percent, and wheat by 130 percent.[117]

The last intervention African people need is to be forced to stop growing their traditional crops that have been nurtured within the climate and soil of the region. But that is what the Gates and Rockefeller foundations are pushing them to do by funding research stations and legal innovations that accept and foster peasant cultivated seed to be transferred from the public good to private market goods. For example, in Zimbabwe, the Gates-funded International Crop Research Institute for the Semi-Tropics has accepted shared seed and then sold the foundation seeds to commercial seed companies without sharing the benefits back to those who cultivated the seed in the first place: "The core of AGRA is biopiracy—the taking of Africa's genetic wealth without benefit-sharing, or even recognition to previous breeders of the seed."[118] While their discourse seems to address concerns of small-scale farms, their financial backing and personnel selected to implement the initiative indicate a propensity toward biotech solutions of new seeds that require intensive chemical inputs more suited to large monocrop operations like the first Green Revolution. Particularly distressing to most African nations, other than South Africa, is their refusal to rule out GM crops. A three-year study backed by fifty-eight governments, entitled the International Assessment of Agricultural Science and Technology for Development (IASTD), completed in 2008 concluded that "GM crops are *unlikely* to play a substantial role in addressing the needs of poor farmers."[119] This report concurs with African farmers that food security for them means replacing the modern corporate model of agriculture with local and traditional know-how practiced by small-scale farmers combined with formal know-how.[120] As La Via Campesina points out, food security is not just some guarantee that every mouth is fed. It must also include who should grow the food, how it is being produced and distributed, who should benefit from the food system, and how to secure a sustainable relationship between the peasant and the consumer.[121] African farmers want to choose what to grow with their own indigenous seeds. African countries have repeatedly stated their desire to maintain the use of their own Indigenous farming practices. Seventy organizations from twelve African countries see AGRA's push for modern laboratory seeds as a Trojan horse designed for "a parasitic corporate-controlled chemical system of agriculture that will feed on Africa's rich biodiversity."[122]

Africans have consistently raised their voices in opposition to GM seeds. GM technology, collectively called recombinant DNA technology, is the introduction of genetic material into organisms that changes the core genetic makeup of organisms in ways that might never evolve naturally. To date there is very little scientific understanding of implications for human health, yet during the George W. Bush administration, the US government ruled that GM crops are the same as hybrid crops and therefore consumers need not be warned when they are in US foods. The *New York Times* reported in May 2012 that "For more than a decade, almost all processed foods in the United States—cereals, snack foods, salad dressings—have contained ingredients from plants whose DNA was manipulated in a laboratory."[123] Typically, GM seeds are genetically engineered to resist pests and can tolerate herbicides, which can increase yields for a few years until the ability to resist pests wears off. The primary concern of African farmers with GM seeds is that they are open pollinators and cannot be prevented from polluting indigenous crops. Within a "few seasons, the GM strain can penetrate well over 50 percent of the plants in an adjacent field, or sometimes one miles away."[124] Not only is the indigenous strain polluted, Monsanto, which controls 90 percent of the global market for GM seeds, sends "field inspectors" into adjacent fields to look for genetic traces of their GM variety. If they find a trace, they then sue the unsuspecting farmer for growing the seed without paying them royalties. So far to date, the courts have ruled in Monsanto's favor.[125]

Suspicions of the Gates Foundation's intentions to pursue GM crops were validated by the recruitment of some top personnel directly from the biotech industry, especially from Monsanto that has been trying to introduce GM crops worldwide since its "Let the Harvest Begin" campaign in 1998. Dr. Robert Horsch, senior program officer for AGRA's Global Development Project worked for Monsanto for twenty-five years. Other AGRA personnel come from the biotech industry, such as Laurence Kent of the Danforth Center, which is heavily funded by Monsanto. In 2009, the Gates Foundation awarded a $5.4 million grant to the Danforth Center to secure approval from African governments for field testing of several GM crops including sorghum and cassava, which are vital drought-resistant and highly nutritious crops for Africa.[126] If these drought-resistant foods were to be polluted by GM varieties, it would be devastating to African agriculture. Adding to the distrust of the Gates Foundation is their refusal to acknowledge their support for GM crops. Dr. Rajiv Shaw, as director of agricultural development for the Gates Foundation and supervisor of both Horsch and Kent, tried to allay concerns by being deliberately vague on their intentions to pursue adoption of GM technology. In 2008, he engaged in listening round tables in the United States to solicit input on GM crops; however, no African farmers were consulted.[127] Shaw first joined the Obama administration as under-secretary of agriculture for research, education, and economics and chief scientist at the USDA. He now heads USAID for the Obama administration. The Gates and Rockefeller foundations have yet to seriously ask African farmers what they want. They continue to hold forums outside of the African continent as they did when they held the third AGRA conference in 2008 in Oslo, Norway. The conference was hosted by YARA International, a global fertilizer company.

The primary way to save seeds is by using them. Seeds can be stored only for a limited amount of time, even if they are refrigerated. If Africans are forced to stop planting their indigenous varieties in favor of a few hybrid seeds or, even worse, GM seeds, the indigenous varieties will not be saved. The predictions of global climate change are for increased drought on the African continent. Without their drought-resistant indigenous crops, Africans will face increased dangers of famine. Additionally, GM seeds are expensive and must be purchased each season along with all of the chemical inputs and intensive water usage we saw in the first Green Revolution. This is not a sustainable model of agriculture.

The preferred model of farming is toward an agriculture based on agroecological principles that is "based on respect for and is in equilibrium with nature, local cultures, and traditional farming knowledge."[128] This model of agriculture is consistent with organic agriculture that has consistently proven to increase the productivity of soil. In the longest study of side-by-side organic and conventional crops, organic crops match conventional yields in good years but organic crops outperform conventional crops under drought and other conditions of environmental stress.[129] "In the developing world, organic yields vastly surpass yields from conventional agriculture by ratios of nearly 1.6 to 4.00. Worldwide across all food stuffs, organic ratios outperform conventional agriculture by 1:3."[130] Earth scientist "Stephan Raspe demands that soil should again be regarded as a 'natural organism,' instead of being treated as a 'lifeless substance.'"[131] A healthy soil is the basis of peasant agriculture.

Who should be able to decide which crops are grown and by whom? The global corporations of *industria* will always make the choices based on their profit motive, not the taste or health of the people or environment. It is clear that, at a minimum, we in the core nations must stop denigrating subsistence farmers as backward and facilitate their ability to continue with this more sustainable way of life.[132] Maybe we could even turn to them for their counsel on how we may choose less destructive ways of life. M. Marcos Terena, a member of the Xane Indigenous group in Brazil, marking the adoption by the UN General Assembly of the Declaration on Indigenous Rights, asks the world to recognize the importance of the contribution of Indigenous diversity to a new world composition of values founded on human rights. He continues, "Such an approach undermines unilateral concepts of development and power, through closer adherence to spiritual and environmental aspects, and mutual respect among people and cultures."[133] It is time to pass the talking stick to those who have maintained alternative values from those promoted by our social prison erected by *industria* and limited justifications based in neoclassical economics.

Chapter Five
A Green Theory of the State

This chapter will attempt to build a new theory of the state. This means the objective is to think seriously about where the nation-state (the state) comes from and what purpose it was designed for, as well as what purpose it fulfills now. The state is the institution that facilitates corporate-led globalization through the penetration and enclosure of the commons by the firm, as we have shown through the Bretton Woods institutions. Understand that when we say "state" or "nation-state," we refer roughly to what people think of as the national governments of countries. However, on the other hand, the idea and apparatus of the state are so much more than this minimal idea. For example, David Held, a scholar of the state, has written,

> The state—or apparatus of "government"—appears to be everywhere, regulating the conditions of our lives from birth registration to death certification. Yet, the nature of the state is hard to grasp. This may seem peculiar for something so pervasive in public and private life, but it is precisely this pervasiveness which makes it difficult to understand. There is nothing more central to political and social theory than the nature of the state, and nothing more contested.[1]

The state is complex, ubiquitous, and inescapably related to imperialism because, while each state may govern itself, where the state has expanded out of Anglo (Canada, the United States, Australia, and the United Kingdom) and continental Europe, the state has reinforced a solid commitment to serve Anglo and European interests. Again, Held writes,

> Whatever its geography, its ethnic complexion, its degree of affluence or impoverishment, the colony's complaint is poignantly the same everywhere: that

its fate is not in its own hands, that its wealth is being drained away to a distant metropolis, that it is made poorer so that others can be rich. The familiar remedy for the complaint among colonies of every sort is economic liberation, securing the freedom to make their own decisions and control their own destinies.... Hard as that freedom has been to achieve, however, it has not been so hard as another kind of liberation—freeing the colonists' minds to imagine fundamental alternatives to the old power relationship.[2]

During the post-war period and during the initial period of the Bretton Woods institutions, a Keynesian approach to economics dominated, which means that the state countered downturns in the economy with state-based spending to ameliorate the worst conditions, including unemployment and underdevelopment. In fact, it is during this period that the academics and agencies from the United States and later other Western industrial democratic nations thought it was their obligation to show poor countries in the global South, usually those countries that had first been colonized, how to "develop." This attempt, labeled "modernization," argued that poor countries could experience the takeoff of economic growth and eventually realize the presumed goal of human destiny, mass consumption, and included the state as the ground floor or vessel that would be filled with modern opportunities.[3] This meant that the state was assumed to tax and pay for public works, such as ports, roads, and education, so that labor could grow around infant industries, which would then develop into revenue strongholds for each country. For example, during this time period, the first part of the Green Revolution was occurring as well, and states such as the Mexican, Philippine, and Indian ones all became fully involved with the research institutes funded by the Rockefeller and Ford foundations and US corporations to develop industrial farming in these poorer countries.[4] As problematic as this paternalistic and Eurocentric approach was to development, the state still at least had a presumed responsibility to the people of that country, and it could not overtly appear to abandon the *public* good—those decisions that would serve the people to make their lives better. As oppressive as this system was, it was nothing compared to the brutality of what would come next—the cowboy capitalism of neoliberalism, which has dominated the approach to economics and governance since it started in the 1970s.[5]

Remember that in 1971, the United States broke from the limits Bretton Woods had put on capital, and the neoliberal era began, along with the growing contemporary conservative movement in the United States, which was instrumental in developing a powerful critique of the Keynesian role of the state, arguing that economic problems were the result of too much interference in the market. Pillars of this contemporary conservative approach, such as Friedrich von Hayek and Milton Friedman, argued that the state's role in governance, economics, and development had to be stripped down.[6] Instead of stripping down the state to create a politics of social egalitarian anarchism, imagined by great thinkers such as Leo Tolstoy, Michael Bakunin, Peter Kropotkin, or Emma Goldman, where the violence of the state is removed to allow for social deliberation about economic boundaries allowing local communities to thrive, the neoliberal order merely stripped down the state where it served the public good and interfered with corporate power more in line with a modified right-wing anarcho-capitalism, or what some have called "jungle law." This all but eliminates any need for the state except for protecting private property (often taken from the commons) and for bailing out large banks and other capital during crises.[7] Social safety nets, education, environmental protection, and redistributive policies to fight economic, racial, and political inequality that were a standard for the modern social contract have all come under fire in the neoliberal world. Here, neoliberalism is a process of destruction where "social relations were being reconstituted in the image of a brutal reading of competitive market imperatives," thus removing the social contract and any recognizable remnant of the public good.[8]

Clearly, the perversities of neoliberalism have affected social mobility, environmental quality, consumption of natural resources, education, social welfare, development, and critical life support systems like food systems. For example, Raj Patel in the academic journal, *Public Library of Science Medicine,* notes, "governments have enabled private sector markets to expand their influence within the food system," while disempowering people and decreasing food sovereignty causing hunger, particularly for women and girls.[9] Patel argues that policies favored by peasant groups like La Via Campesina would address fatal power asymmetries that distort the food system in unsustainable ways. The neoliberal transfer of power meant that the state should not regulate as much—it should not strenuously regulate banking and finance, manufacturing, or extractive industries that would grow much faster without this government monkey on their back, and this would keep economic prosperity growing in the countries that adopted such an approach. All the while, as corporations are loosed from public restraints, the public has less input for redress of grievance while food, energy, and politics become increasingly hierarchical, despite the shadows of democratic pretenders in nationalistic displays, such as during elections, but leave little discussion about real political options outside of neoliberalism. None of this is to say that the state is necessarily less powerful than corporations, but that the state has demurred purposefully to corporations in an alliance that has little pragmatic or substantial accountability to sustainability or inequality. Today, the state has become the most important handmaiden of corporations who have pursued profit and power that ravage the planet while dissent is suppressed or trivialized. Corporations derive their profit from unhindered access to planet Earth's resources facilitated by the state, yet there has been no clear theoretical explanation, no theory of the state, of how or why the modern state fulfills this function to the exclusion of almost all other interests. However, this process did not appear magically or without a history.

Simply, the trajectory of modernity has been one of shifting control and power, particularly over markets and economic decision-making. All people had been tribal in the past, where power was deliberative and generally noncoercive, because tribes typically did not claim the power to control individuals, rather economic decision-making was based on consensus and collective decisions—the social group totally determined economic relations.[10] Then, as populations grew, hierarchy did as well. Karl Polanyi notes that during mercantile capitalism in the colonial era, the economic system was controlled and was secondary to culture, family, and village or town. But, the "great transformation" to market capitalism in the nineteenth century led the economic system to be controlled by the market.[11] He notes this fundamental shift from social concern and focus, to an economism that places economic values above all else:

> Ultimately, that is why the control of the economic system by the market is of overwhelming consequence to the whole organization of society: it means no less than the running of society as an adjunct to the market.... The vital importance of the economic factor to the existence of society precludes any other result.[12]

Crucial to these modern hierarchical changes has been the role of the state. Here we will draw a more relevant theory of the state that explains the state's ecological and cultural purpose has been curiously missing. Most theories explaining the development and purpose of the state focus on divisions of labor and controlling for the misery that ensues as property and status are conferred on some, to the neglect of others. The state, it seems in the mainstream literature, is an instrument of control—this much is clear. But why? For what purpose? Does it appear magically as in the structural functionalist sociology theories? Does the state grow from the "state of nature" as a rationally produced force of stability and the rule of law? Is it the product of capitalist elite, manipulating legitimacy and consciousness to avoid a revolution of the capitalist system? These are the questions the three primary theories of the state ask—but where is the power that stems from organizing nature in all of this?

We will first locate the state's ecological purpose within the framework of Hipwell's *industria* discussed in prior chapters.

Industria

Hipwell notes that *industria* is a predatory world-system of knowledge and power:

> In the view of many scholars and activists, the political, economic and ecological phenomena described in the preceding pages are the effects of a single but

multiplicitous system of power/knowledge. According to this view, the diverse empires and state alliances that have historically struggled amongst themselves for planetary dominance have finally merged and accreted into a single, self-regulating and self-perpetuating mechanic assemblage.[13]

Hipwell explains that *industria* is a homogenizing force, "a network not a container," the opposite of wilderness, expansionary, and finally, a system of knowledge and power, not an ideology.

As in Deleuze and Guattari, who refer to "Urstaat" and the "state apparatus," which Hipwell notes confuses individual states and the state system and the state-corporate network—all better explained in the term *"industria"*—is focused on territorial and social conquest and "actively works to impose the order and calculability of systems on manifestations of difference, such as diverse cultural groups or wild ecosystems."[14] Thus, it is designed for "striation, order and domestication," which strip value from diverse cultures and ecosystems to funnel value/energy to the core. The network operates more from core cities as the important nodes rather than countries. "Since industrial civilization is a web of geographical power completely encircling the Earth, there can be no linear, terrestrial frontier delineating its furthest reach. Its 'outside' or 'periphery' is found in the smooth spaces between the strands of the web."[15]

These spaces between the strands of power, connecting nodes of *industria*, are the geographic places where *industria* and its homogenization are absent enough to allow other "wild" species to exist, for example, timber wolves, grizzly bears, or mountain lions. These wild spaces are not left to be on their own, however, as *industria* is a predatory and expansionary force, constantly changing and growing to annex areas, such as the hinterland surrounding municipal areas, as well as the last remaining wild spaces.

Finally, this system is a network, not an ideology, since it is found in capitalist, socialist, fascist, and other regimes. However, *industria* at this point is "almost universally capitalist."[16] Thus, Hipwell rightly describes the state within *industria* as one that is expansionary, insatiable, and bent on organizing diverse cultures and nonhuman nature for the purposes of consumption. Our vision, and Hipwell's, is consistent with the profound work of James C. Scott, who has documented the homogenizing and consumptive force of the state in Southeast Asia.[17] Scott has shown that the state has a history of attempting to homogenize and simplify diverse cultures and nature to make people and nature easier to control. After decades of research on the consequences of "seeing like a state," he has boldly asked us to take up the left-wing anarchism (right-wing anarchism is anarcho-capitalism) of social and voluntary cooperation that would not allow for the wanton yet deliberate destruction that comes from the institutionalized hierarchy of the state.[18]

Yet, theories of the state currently lack a sense of how important this organizing of nonhuman nature is to the state. In creating a green theory of the state, we are attempting to position the control of nonhuman nature as the central purpose of the state apparatus. To do this, we suggest that the state must be viewed from a perspective that highlights its social construction. This approach pointedly avoids "naturalizing" and normalizing the state by pretending the state was not created by people but has always been there. This occurs in Hobbesian mythology when Thomas Hobbes believes there was a time when people elected to leave an anarchic state of nature to enter into a social contract. But, because the state of nature itself is a myth, Hobbesian ideals mystify the state's origin and purpose. We suggest that Indigenous politics is a lens that can lend this perspective given that a core concern in Indigenous politics is current-day decolonization at a global scale and much of this is situated in state-Indigenous relations.[19] We saw some of the inherent dynamics involved in the Zapatistas' struggle for autonomy against the Mexican state in Chapter 4. This autonomy included a moral right to their own identity that could not be achieved without control of their communally owned land.

Indigenous Insight on the State's Domination of Nature

Looking through the lens of Indigenous peoples' experiences, the state can be clearly recognized as a socially constructed institution oriented toward the imperial purpose of dominating nature and all those peoples associated with it. David Maybury-Lewis, who is both an Indigenous person as well as a renowned scholar in Indigenous studies, writes, "The killing of indigenous peoples is usually resorted to when outsiders wish to seize the lands and resources they control or when the indigenous populations are simply considered to be 'in the way' of national destiny, development, natural resource extraction, dam building or anything else."[20]

In order for these genocidal activities to have been possible in the first place, Europeans needed to command, extract, and order their own natural resources to create the weapons and transportation and wealth needed to sail to distant lands and slaughter the people they discovered there. Dams are not just built. No—the managerial, martial, and knowledge-based power needed to build something like an industrial dam is, so far in human history, based in the state. And, build dams we have, but the repercussions may be catching up to the First World and *industria*.

Indigenous peoples are also in the forefront of warning us that Mother Earth can no longer survive this domination and destruction of nature by the state. For example, the Hopi tell us that we are in the fourth incarnation of the world, which has followed on the heels of collapse and apocalypse of three other worlds prompted by greed, avarice, violence, and self-involvement.[21] Today, the Hopi caution that humans are still responding to these same destructive impulses that have pushed this Fourth World into its last throes in which it is literally being devoured. Hopis are not the only ones noticing that we are in trouble. Collapse is not in the future, it is occurring *right now*. Though most people are aware of the extinction of the dinosaurs, biologists tell us that we are really in the throes of the Sixth Great Extinction of life on Earth, and this one has been underestimated and is picking up speed.[22] Life on Earth is being lost, species extinguished at 100 to 1,000 times the normal rate of extinctions, and it is human activity—especially our conversion of land—that has taken habitat from the rest of creation and pressed plants and animals into the recesses of token spaces to survive.

We have fundamentally altered the chemistry of the oceans, soil, and atmosphere, and have done so in short order, principally in the name of industrial growth and development. But, this carnival is not centered everywhere equally: no. This insatiable and predatory system of knowledge and power found in the current network of global firms, military apparatus, states, and the "elites who control them" was centered first in Europe during the modern period, and now in the United States.

The fact that the industrial world is unsustainable is hidden in complex political economic arrangements[23] that allow most of us in the core areas of *industria* to go to school, work, and enjoy leisure without seeing or confronting the unraveling of ecosystems. The carnival has such pretty lights and hypnotic rhythms, and the cadence of the carnival barker is almost irresistible. But, life on Earth depends on our peering beneath the surface and even resisting the waves of force that *industria* will unleash when it realizes we are not deceived anymore.

We argue that the accumulation of power found in early European societies was based on an agricultural economy that was organized through the development of key political institutions, namely, the state, the firm, and the current financial international governmental organizations that govern corporate-led economic globalization. All of these institutions had and have as their foremost goal to accumulate wealth and power—but what has been neglected is that they have done so through the decision to command and organize nature that has been enabled by the hierarchy and central apparatus of the modern state. The state is a European concept and organization that has now been fully globalized into a regulating system of states with nearly full governing authority around the world. It was the organization and command of nature through agriculture, as well as timber harvests and some mining that allowed the states to accumulate sufficient wealth and power necessary to set up their hierarchies. Thus, the modern state developed as the core benefactor of exploiting nature, particularly agricultural surplus early in human history. As Wilmer says, "The First World is made up of the political communities which were the first to organize politically as nation-states, organize economically to create surplus agricultural capacity followed by industrialization, and develop democratic political systems."[24] The nascent states gained power, Wallerstein writes, through four developments, "bureaucratization, monopolization of force, creation of legitimacy, and homogenization of the subject population."[25] Further, the machinery of the state was necessary for the development of the capitalist world-system, which develops alongside the European state apparatus first embodied in absolute monarchs of western Europe: "It is evident that the rise of the absolute monarchy in western Europe is coordinate in time with the emergence of a European world-economy," where the "state structures were themselves a major economic underpinning of the new capitalist system (not to speak of its political guarantee)."[26]

This added power allowed for a growing population to diversify and create a division of labor that included professional soldiers and leaders that did not have to provide for themselves.[27] By the end of the feudal period such accumulation of power allowed these Western centers of agriculture to extend out into the world and draw natural capital out from peripheral areas in order to feed their growing need for land and raw materials.

The state as an institution is key to this process because it concentrates managerial, material, ideological, and knowledge-based power within an apparatus with a high degree of hierarchical, and therefore ordered, organization that can manage all of the complex and resistant elements within and outside of that society. The organizing of these various elements allows for power to be directed to elites in the Western societies who then were able to reinforce the process of concentration and expansion.

It is critical to examine the role of the state now because the few remaining public spaces—the oceans, the atmosphere, the biosphere, all fundamental to the survival of the planet—are under threat of collapse. The state has thus far been a handmaiden to corporations in exploiting and penetrating further into the commons of the Earth, but the state is also the only institution powerful enough to ever challenge the firm.[28] Once the state is no longer as necessary, its prominence, as is already evident, will shrink to allow the power of the firm (a private form of the state) to take its place so that the last remaining capital concentrated in the public sphere can be transferred to the private sphere. In the end, *industria*'s goal is to accumulate as much wealth and power as possible under the control of private elites at the center of the system. Each institution that is prominent in world politics today appears to reflect this process in terms of first the state, then the state-based international financial organizations, particularly the Bretton Woods institutions, and then finally the transnational firm (which almost always is reaching into the periphery, but firms from the periphery are almost never found to be doing the converse). In Chapter 6, we will look to Indigenous people's resistance to state power as a guide to understanding alternative roles for the state, but first we need to understand the historical formation of this destructive state role and why dominant theories of the state have failed to provide a corrective.

Social Construction of the Modern European State

This section of the chapter will briefly look over the dominant theories (sociological, liberal, and Marxist) of the state to assess the prominent position the state has in political theory, and at the same time identify the Eurocentric and modern culture that the state fundamentally represents.[29] We will show that the state's role has been identified as one of authoritative Western global control but leaves nature out of the understanding, and is therefore incomplete.

The state is the "apparatus" of national government, but it is also so much more.[30] Held notes that the state is a part of politics that is everywhere and nowhere at the same time, because its power is felt throughout the most personal levels of society, but is also very difficult to tangibly locate. The apparatus of government here at a minimum level is the central government of a country, but the state infiltrates public and private moments, rules, and opportunities that are not recognizable in the simple form of a national law or a national military or police force. Apparatus here also implies a system of social control that extends deep into a society through social networks and, as Antonio Gramsci would describe in his prison notebooks, this network of social control is hegemonic, but covertly so.[31] In other words, the apparatus of the state is like an iceberg, or perhaps like a cockroach on the kitchen counter—what you see is only a fraction of what is there. Gramsci describes this hegemony as a class-based state apparatus that reaches down into schools, churches, markets, and families. Here, control is exerted through the expectations and social demands to, simply, not challenge the hegemonic power. For the most part, citizens become trained, and in the case of nationalism extremely so, to not only refuse to question the central foundations of the state as an institution, but to defend it through national pride.

Our ideas of the state have been extraordinarily limited, and few reach the criticisms or scrutiny of Gramsci who sees the state as a latticework of hegemony and repression working in defense of class and dominant cultural interests.[32] Instead, most understandings of the state are apologetic at best, and naturalizing at worst to the extent that the state is assumed to be a natural extension of civilized societies—even though the state began as almost exclusively a western European institution. Sociological theories, such as those of Auguste Comte,

Emile Durkheim, or Max Weber, see the state as a naturally evolving institution meant to serve an increasingly complex society. Durkheim is among the most important thinkers regarding the state. As a founder of the "functionalist" school of sociology, he likens the institutions of society to an organism. This school of thought uses the organism for two reasons. It not only sees institutions as developing "naturally," it also posits that these institutions are meant to work together for the function of the whole organism. This coordination and control become important for sociological theories that see several pathogenic results of complex society and divisions of labor that hurt some people.

Indeed, legitimacy of the state's use of violence was articulated by foundational sociologists such as Durkheim and Comte who suggest that the state was the source of modern and moral authority.[33] Weber further elaborated the point when he noted that like all political institutions, the state "is a relation of men dominating men," but this domination serves as a "*monopoly of legitimate use of physical force* within a given territory."[34] Because the state is the source of legitimacy, its actions are not recognized as violent. Or, if they are seen as violent, the violence is not seen as problematic as it furthers the goals of a social contract and modern progress. This is also reflected in the roots of Georg Hegel's *Philosophy of Right*, where the state is the ideological foundation of a good and developing society, an idea Marx would later dismiss.[35] For example, the state's imagined monopoly of the legitimate use of force was the justification for the violence used against American Indians in the establishment and maintenance of the United States. Taking and controlling tribal land was merely seen as a necessary step in the progress of the state.[36] Thus, the classical sociological theories of the state indicate that the state is an element of social evolution that is "natural," that is, not politically constructed but organically formed without anyone really making a specific decision to install a regime.

Under liberal theory, the state is constructed as a force for limiting human violence through a social contract for domestic policing and the post-Westphalian politics for international policing.[37] Since early discussion of the subject by Greek scholars, for an action to be considered violent, it must be an illegitimate, irrational behavior of a minority of individuals in society.[38] During the casting of modernity, this became an axiom of the social contract. One purpose of the social contract was to keep violence to a minimum so that people could be free to live their lives without the risk of violence that was thought to exist outside formal social organization. According to Jean-Jacques Rousseau, to enter the social contract is to gain civility and the ability to have real property. "What man loses by the social contract is his natural liberty and an unlimited right to everything he tries to get and succeeds in getting; what he gains is civil liberty and the proprietorship of all he possesses."[39] Thus, the social contract provides the supposed civility of imposed limits on violent human appetites while providing a system where the possession of real property is possible. Outside the social contract, at least to the original social contract theorist, Thomas Hobbes, there is no such thing as "private property," merely the ability to temporarily use a resource.[40] This is important because the doctrine of discovery allotted private property rights to "discoverers" and only use rights to others, such as Indigenous peoples, who were seen to be outside the social contract, residing in an uncivilized "state of nature."

Outside the social contract and within the anarchic state of nature, violence is an expected behavior, and for this reason, it was a rational decision to sacrifice some portion of individual liberty to a sovereign who would keep order.[41] Thus, violence within modernity is usually conceived of as the erratic behavior of criminals and has not typically been conceived of as a social continuity perpetrated by rational and civil modernity itself. By definition, the state becomes a protector from violence, not the perpetrator of violence; and, violence that the state does commit is veiled in legitimacy. Thus, the state is the refuge from the brutality of nature as it is the embodiment of modern civilization, where property and industry may progress—and even become the ideal of progress.

Thus, in liberal theory, the purpose of the state is to impose an order that quite literally allows for the transition of "nature" into "civilized" property. Note that nature here is quite worthless without the institution of private property and the state. Importantly, the idea of drawing Indigenous peoples into "civilization" has consistently been framed as something for their own enlightenment and salvation.[42] But from this reading we can see that what Europeans really wanted was to convert Indigenous and nonhuman nations (what Europeans saw as "nature") into commodified property. This funneled value directly to the elites of the state—such as the landed aristocratic class or the bureaucracy itself—as the European states moved outward.

Marxist theories of the state, which are less an elaboration of Marx than of later scholars, also see the state in terms of control, though this control is in the interest of the capitalist class. Held explains that there are two central Marxist visions of the state:

> The first, henceforth referred to as position (1), stresses that the state generally, and bureaucratic institutions in particular, may take a variety of forms and constitute a source of power which need not be directly linked to the interests, or be under the unambiguous control of, the dominant class in the short term. By this account, the state retains a degree of power independent of this class: its institutional forms and operational dynamics cannot be inferred directly from the configuration of class forces—they are "relatively autonomous." The second strand, position (2), is without doubt the dominant one in his writings: the state and its bureaucracy are class instruments which emerged to coordinate a divided society in the interests of the ruling class.[43]

The state in Marxist theories is a "net" of institutions with relative autonomy that both serves and negotiates with the capitalist class and is used to maintain privilege and protect the bourgeoisie from an unhappy working class.[44] In later neo-Marxist theories, the way the state serves capital became more clearly articulated. One school of thought, the "instrumentalist," argues that the people who fill the ranks of the central government personally advance the interests of capital, and another argues that institutions have their own power and, regardless of the personalities, will serve the interests of capital.[45] In particular, Marx noted that the state, through its network, could pass information easily and had the capacity to neutralize or undermine social movements that threatened the status quo.[46] Importantly, the "repressive dimension of the state is complemented by its capacity to sustain belief in the inviolability of existing arrangements."[47] In other words, the state protects and negotiates with elites, represses social configurations against these interests, and then makes the circumstance appear inevitable, natural, and normal—beyond question.[48]

In this way, the Marxist theories of the state peer more closely into the construction of power, including the secrecy and mysticism that permit such accumulation of power such as occurred in the repressive class politics in London during the nineteenth century. Sociological and liberal theories both work to reinforce the state's own desire to seem inviolable, and thereby reinforce the power of the state itself. Consequently, it appears to us that the Marxist theories come the closest in the triad of dominant theories to understanding the diffusive yet hierarchical power that the state holds and what this means for nonhuman nature and those resistant to *industria*.

Yet, Marxist theories of the state, while they describe alienation and objectification, also neglect nonhuman nature as an absolutely necessary force and target of the state. And, while the Marxist theories of imperialism are relevant, none of this, as a theory of the state, helps explain why or how the same London matrix of power was able to extend itself across the World Ocean and, with other Europeans, command much of what they beached upon two hundred years prior to Marx's arrival. Marxist theories note that the state's purpose is to serve an elite and placate/suppress/undermine the working class, but there is little connection between this and the necessary exploitation of nonhuman nature to make this effective.

Mainstream Observations

Importantly, the sociological, liberal, and Marxist ideas of the state are similar in several ways. First, all three of the triad are Western theories. That is, they all take aspects of modernity and the Enlightenment for granted, such as the notion of a progression or stages of society, and at the same time, they are the dominant modes of literary explanation for this central feature of life we call the state. Other theories that call into question the legitimacy of the state in general, as in anarchism, are fully marginalized in political science, and notions of Indigenous worldviews in contrast to the state are all but fully ignored. Part and parcel with this modernism comes a vision of human destiny that sees the state as necessary and even natural. Even in Marxism, the state is a force under capitalism and socialism, and only under Communism does it fall away. But without those prior stages, Communism is a distant leap.

Also, all of these ideas see the division of labor in society as a purposeful benchmark for the establishment of a modern state. Even though Marxists see this division of labor as a core social problem, the other two approaches see the division of labor in society as a natural requirement for more and better production, and therefore it is a beneficial element. This is likely an artifact of the modern era itself that witnessed how the industrial transitions complicated divisions of labor in society and destroyed traditional divisions.

In each theory, the state becomes more pronounced and necessary as labor specialties and work become more distinct. Society becomes more alienated from its social core affinities and more individualized, and Durkheim warned that if this individualization and division of labor developed too much it would become pathological.[49] However, the purpose of the state was to stand back and allow for this division of labor to occur, or from the Marxist critique, to enforce this division of labor with false ideology and coercion. In any case, the modern industrial division of labor is of elemental concern because it becomes one of the central reasons for the state to exist. In the functionalist perspective, the state comes to exist through a natural need to sew the various parts of the organism together. In the liberal form, the state comes to exist to enforce private property that comes from labor, and its division allows for more affluence, property, and expansion.

But where is nature? None of the power that is wielded by the early imperial states is possible without organizing nature *first*. In order to do this, a strong hierarchical regime is necessary to command the seizure of local nonhuman nature to build the initial power base—as the British did in converting their forests into ships—and then using this power base to pursue more nonhuman nature and peoples to funnel into the machine that converts this all into treadmills of wealth and power. This is evident when the British used those ships to sail to distant lands and commandeer peoples' lives and natural wealth in what would become their empire and, today, the periphery of *industria*.

A Green Theory of the State

Much work has been done on understanding the state as an institution, so why another theory? The mainstream theories—sociological, liberal, and Marxist—all are within the modern tradition, focus primarily on the issue of social divisions of labor and authority, and almost totally neglect the relation of the state to nature as grounding for Western power. Thus, a green theory of the state is necessary and appropriate because current theories of the state neglect the way that the state has been *designed* to command nature in order to command the world, thereby ensuring a flow of material power to the hegemonic elite within the world-system. This is important for green theory in general, because one of its central purposes is to explain ecological change and destruction. A green theory of the state contributes to our understanding of how the state acts as an institutional channel for predatory international consumption by organizing nature and related populations for the benefit of social elite.

What constitutes a theory of the state?

If we are to create a new theory of the state, what should it include? As Held notes, understanding the state is a preeminent but almost ethereal project in political science. Nonetheless, there are a few elements that are necessary to a theory of the state. While a theory of the state should describe the governmental structure found in society, it should also go beyond identifying a regime type, bureaucratic orders, and the formal rules found in levels of government. Government also possesses a character that is composed of the more pervasive and less visible areas of state structure. Note that we are using the term "character" of the state instead of "nature" of the state since "nature" invokes an image of permanence, propriety, and normality that we do not intend. "Character" here means that the state has a personality, which is both visible and invisible in its exercise of power. This personality is filled with motives and actions that are directed by people, even if these people create structures of power, which later run autonomously as in the structural neo-Marxist argument made by Nicos Poulantzas.[50] On the other hand, this is not to say that the people within state identities work with complete agency, rather that the structures that form state identities were at one time formed by individuals and groups. More importantly, these groups and individuals institutionalized knowledge and power systems into the state, and these institutions are keys to identifying the character of a state.

In order to explain the way change occurs, a theory of the state also needs to include an accounting of how civil society operates with or within the state. Fundamentally, this requires examining the agency of the private sphere. Is civil society independent of the structures of the state? Is it completely under the thumb of ideological hegemony as described by Antonio Gramsci,[51] or is there some sort of dialectic between structural state systems and civil society agency as described by Anthony Giddens?[52]

In addition to accounting for civil society's agency (or the repression thereof), a theory of the state should also account for class and other power structures. This means that theories of the state should typically include at least a limited political economy.

In sum, a theory of the state includes an accounting of

- formal and informal governing institutions and their operating knowledge power systems;
- a description and characterization of agency within the private civil sphere; and
- a political economy that is part of the above.

What has been done in relation to a green theory of the state?

Whereas there has been a great deal of thinking about *how* to green the state, a "green theory of the state" is not yet cogently established in the literature. For example, Robyn Eckersley has written on several occasions regarding a green theory of the state, by which she means a "greening" of the state. In *The Green State,* Eckersley argues for a "post-liberal" ecological democracy. Her plea is that "all those potentially affected by a risk should have some meaningful opportunity to participate or otherwise be represented in the making of the policies or decisions that generate the risk," and those that cannot be present, like future generations and animals, must have their interests treated *as if* they were.[53]

Importantly, Eckersley notes that states are here to stay and must be modified to be just and sustainable. The same kind of project is in place in her co-edited volume with John Barry, *The State and the Global Ecological Crisis,* and is in related works from John Dryzek et al.[54] Yet, as important as it is to democratize the state in pragmatic terms, these projects do not establish a green theory of the state in the sense that the theories explain the origin of the state or how and why the institution functions as it does. Further, Frank, Hironaka, and Schofer[55] note that environmental protection has become a "basic state responsibility." Throughout world society, it is expected that states have the responsibility to protect environmental conditions. Thus, the role the state has assumed in relation to nature, according to Frank et al., is its *protection*. Frank et al. are not working within green theory per se, but these efforts are all similar in that they theorize the state's *potential* in nature protection and pursuing sustainability. Eckersley acknowledges the tensions between the state's history as an institution and its potential, but says that the state is the only institution with enough power to work against firms and other key power brokers to make sustainable political action a reality.[56]

Murray Bookchin has approached some of the sentiments of a green theory of the state. He notes, for example, that the state is necessary for the domination of nature:

> It remains one of the most widely accepted notions, from classical times to the present, that human freedom from the "domination of man by nature" entails the domination of human by human as the earliest means of production and the use of human beings as instruments for harnessing the natural world. Hence, in order to harness the natural world, it has been argued for ages, it is necessary to harness human beings as well, in the form of slaves, serfs, and workers.[57]

Bookchin's anarchist social ecology is necessary in order to resist hierarchy and domination of nature by the state. From here, Bookchin carves out the notion that governance should occur via a confederacy of municipalities. However, if we look back at the notion of *industria,* municipalities are still nodes of central control. This control is related to Barry's observation that Bookchin's confederacy still resembles the state to some degree:

> One can interpret Bookchin's argument for devolving power to municipal levels, yet maintaining a legitimate right for the confederal council to intervene in municipal affairs, as bestowing state-like institutionalized powers on the council.... From this it is not stretching things too far to suggest that this agreement functions as a sort of "ecological social contract," which on familiar contractarian grounds legitimizes the state.[58]

Also, some work has tangentially been done by Matthew Paterson in his case of sea defenses in Eastbourne, United Kingdom, where he notes that "nation-building" is a part of the "specifically modern twin projects of the human domination of nature and of some humans by others, and in the dynamics of capital accumulation, or 'economic development.'"[59] Here Paterson posits that nation-building is "simultaneously a part of the human domination of nature, of the imposition of human physical control over land and water."[60]

Paterson's claims are similar to Worster's in *Rivers of Empire,* where he notes that there are two important kinds of states—agricultural and capitalist states where societies built around the control of water (hydrologic societies) emerge as a result of the control of nature under conditions that require a high degree of capital, organization, and power.[61] Thus, Bookchin, Paterson, and Worster all indicate that the state, as an institution, has developed around the need to control and manage—indeed dominate nature. However, Paterson's treatment is not specifically meant to explain the emergence of the state but rather to position the state in terms of domination. And Worster's thesis is limited to hydrologic societies, or those that become dependent specifically on an elaborate and technical system of water control through irrigation or commercial development, usually in arid societies, and Worster does not comment on the hydrologic state within a system of states but instead treats the development of these states in isolation.

In sum, these authors document that the structures of the state are the only ones powerful enough to protect the commons, such as the World Ocean or global atmosphere, and that they are necessary for planetary survival and that world society expects this responsibility to be filled by the state. However, it is also clear the state has actually been used as a tool for the domination of nature. In order to explain how this path of domination has been chosen we must first go to the foundation of the modern states, the absolutist European monarchies.

Some essential elements to a green theory of the state

Work still needs to be done to really establish a green theory of the state; this section will begin this process and attempt to fill in the requirements of a true theory of the state. The basic proposition is that the state was devised to control nature and people, to create a growth-oriented political economy that serves the most powerful consumptive classes in the world-system, and that civil society is imperfectly constrained within the opportunities that the ideology of growth and its world capitalist system provide—but that this system is creating contradictions that are fatal to the system itself. One such contradiction is the exhaustion of environmental systems that feed material economic growth, and another is the elimination of biological and cultural diversity that are the base of the Green Revolution. The crises that these contradictions create may provide windows of opportunity to change the world-system.

According to Held et al., the ancient powerful societies were not necessarily states—they were empires of conquest (e.g., the Chinese, Japanese, and Islamic empires). They did not establish governance, lacked administrative and local control, had little in the way of lawmaking and enforcement, and had extremely amorphous boundaries that waxed and waned at the frontier on a regular basis with conflicts between other empires. These apparatuses of the state, in addition to standing armies and a homogenization of societies *within* the polities, did not begin to emerge until the establishment of absolute monarchies of Europe and the Italian city-states of the 1400s.[62] "From the outset this process involved great costs for the autonomy and independence of many, especially in non-European civilizations. In fact, the spread of the interstate system has been consistently characterized by both hierarchy and unevenness as Europe burst outwards across the world."[63]

To be clear, some of these monarchies became empires, but not all empires had the apparatus of the state behind them as did the latter European empires.

This "burst outward" was something new in that it now included and exported a "centralization of political power, the expansion of state administration, territorial rule, the diplomatic system, the emergence of regular, standing armies—which existed in Europe in embryo in the sixteenth and seventeenth centuries [and] were to become prevalent features of the global order."[64] Indeed, between 1400 and 1600, European feudal principalities had been consolidated, and the birth of the state took form.[65] Thus we know that the state as a modern institution is built on an expanding hierarchically "administered" life that was distinctly European. The apparatuses of the state were socially constructed by those benefiting from this system in a dialectical process between the elite of civil society and the state.

But why? Why did the European state "burst outwards?" Held et al. note, "While the process of equipping, planning and financing overseas exploration sapped national resources, as did the management of newly acquired posts and territories, governments reaped some of the *fruits* of the 'discovery' and exploitation of non-European lands."[66] This "fruit" fed further expansion of the European state system that became the "rapidly developing empires of Britain and of other European states" that were "the most powerful agents of globalization in the late nineteenth century."[67] The continued expansions of the European empires depended upon the resources they found in these peripheral areas to continue to fund further colonization. For example, "Exploitation and export of gold from the Antilles supplied necessary capital to the Spanish empire for other expansionist efforts and was essential to Spain's power at the time."[68]

This capacity then allowed for a deepening control over natural resources and Indigenous people, as seen in the infamous Slave-Sugar-Cotton triangle that required the state to dominate the oceans in transatlantic shipping. In this case, the French, English, and Dutch empires worked in concert to ship slaves from the African Continent to the Caribbean and North America to tend to agricultural crops such as sugar and cotton, which were then shipped to Europe for trade for manufactured goods.[69]

Indeed, in each of the cases as European empires established themselves around the world, they appropriated through violence, coercion, and co-option the natural resources of the area and transplanted a brand of the European institution meant to enable the ongoing core-periphery relationship, even after some of the physical colonialism ended in the twentieth century for most areas in the Third World.[70] Thus, the European state burst out in order to control and direct nature from overseas back to the monarchies to further fuel their ascent to world power. Gold bullion was important, but even more central were the staples—agricultural goods.[71] In order to establish control upon colonies, the European state used its military apparatus to impose the European administrative state. Within these newly acquired territories, the settlers bestowed power and privileges to a minority of locals who then developed an interest in cooperating and maintaining the imperial system. This system proceeded to arbitrarily draw borders, imposed administrative rule of law over local and traditional forms of Indigenous governance, organized and controlled labor, and ensured that the cargo of resources flowed back to the European continent.

Our proposition therefore is that the formal institutions of government, at first the absolute monarchies of Europe, which consolidated governing power into a new centralized administrative state, were set up to favor the central elite of the world-system. This institution then put in place an international capitalist political economy that ensured the flow of natural resources to the core powers of Europe both through the first states in Europe, and then through the establishment of colonies and later the development of new states in Latin America, Africa, Asia, and Oceania. The knowledge/power system was mobilized by an elite portion of civil society in pursuit of a capitalist political economy to set in place mechanisms of hierarchical control embodied in the newly emerged administrative apparatus of the modern state.

What of civil society and agency? If the idea of citizenship is tied to the idea of the state,[72] then civil society develops as a result of the state and the resulting political economy. In the dialectical process, which established civil society at the origins of the modern state, agency was restricted to an elite. While the elite may not have possessed complete agency to dictate the exact goals of the modern state, they were vital to a capitalist political economy that the European states saw as their route to accumulation of increased state power. Those not aligned with the knowledge/power system set in motion to support this political economy were suppressed. Thus, the potential for civil society, which included all members developing agency to contest this privileging of the elite, was smothered by state repression. The personality or character of the state that forms the visible and invisible exercise of power was socially constructed and institutionalized by those benefiting from the status quo.

The International System of States

The above sets out how modern Western society has developed using the state as an agent of command and control of nature and people. However, we now live in a world where there is a global system of states, not just European empires. How did the establishment of these empires affect the state system we have today? Control of world resources allowed for the continued political power for core states, and the loss of ecology drained the self-determination of Indigenous and peripheral peoples.

Stephen Bunker and Paul Ciccantell explain how core states maintained political power through the process of ecological unequal exchange:

The manipulation and reorganization of the relationship between nature and society is the most complex task confronting any ascendant economy. Gaining secure, inexpensive access to the huge volume of raw materials building blocks of capitalist industrial production requires economic, political, technical, and organizational innovations that restructure both existing social relationships (e.g., core-periphery relations) and the characteristics of the nature-society nexus (e.g., what raw materials are extracted and by whom). The strategies of states and firms in ascendant economies to accomplish this task create what we term "generative sectors": leading economic sectors that are simultaneously key centers of capital accumulation, bases for a series of linked industries, sources of technological and organizational innovations that spread to other sectors, and models for firms and for state-firm relations in other sectors. These generative sectors in raw materials and transport industries have driven economic ascent throughout the history of the capitalist world-economy.[73]

In other words, the state develops sectors and extensions that feed the core states and firms that profit from this system. This relates to Bunker's idea found in *Underdeveloping the Amazon*. In this seminal work, the concept of "ecological unequal exchange" is developed as an extension and more profound explanation of dependency theory, while also adding several key modifications.

First, Bunker sets out Nicholas Goegescu-Roegen's thesis of "entropy and the ecological problem" where the laws of thermodynamics help explain international development. Entropy is the disorganization of matter and energy. Any human use of ecology depletes the useful energy of that ecology. If nature is a storehouse of energy, then the most valuable aspect of this storehouse is when it is intact, and this value then decreases over time as ecologies are harvested and exploited through the production of commodity chains.

Energy contained in natural resources, which is essential for the development of social complexity and power, is transferred to the core areas of production. These raw materials, like timber, are valued minimally in the human economy, even though they contain the greatest amount of well-organized energy (least entropy) that they will ever have along the production chain. This production chain systematically erodes the energy and speeds up the entropy of the energy found in the tree (say, as it turns into a desk, or pressed wood fibers or matchsticks). The energy from ecology is converted into economic value through labor and processing done in core areas. Thus, ecological unequal exchange starts with the notions of energy and entropy and explains that the core areas in the world-system are more affluent, economically and institutionally complex, and quite simply more industrialized because they are "feeding" off of energy that comes from somewhere else. However, that "somewhere else" (the periphery) is being systematically eroded in its capacity to become affluent because the energy needed to do so is going to the core. As the core becomes more predatory, nature and its well-organized energy with low entropy becomes the target of a very efficient and increasingly powerful system. Yet, this destruction of nature in the periphery ultimately threatens the core also because as nature becomes less well-organized (contains less useful energy), the entire expanding modern system becomes unsustainable. *Industria* is setting itself and the rest of planet Earth up for social collapse, even as it undermines the periphery's ability to sustain itself. *Industria*, then, is at once simplifying the social and natural systems of the world toward one that has fewer and fewer sources of usable natural capital in its pursuit of accumulating wealth and industrial power that are needed to climb the steeper and steeper slope of gathering energy and nature—consuming the very capacity to foster its own metabolism in the end. This is one of the contradictions of the current system that threatens the status quo, mentioned above.

However, the core will be, even in this model, the last to starve, in addition to the fact that in this model, the core requires the periphery to starve as they consume more, assuming a relatively limited amount of natural energy.

Bunker explains,

> The flow of energy from extractive to productive economies reduces the complexity and power of the first [periphery] and increases the complexity and power in the second [core]. The actions and characteristics of modern states and of their complex and costly bureaucracies accelerate these sequences. Modernization, as ideology, as bureaucratic structure and procedure, and as centralized control through complex regulatory organization, mediates and intensifies the socioeconomic consequences of the interaction between global and regional systems.[74]

Further,

> Modern systems are themselves highly energy-intensive and can only emerge in regions where industrial modes of production derive large amounts of energy and matter from subordinate modes of extraction.[75]

Bunker notes that a partial understanding of development is found in this zero-sum relationship that determines the distribution of energy and power, iterated and institutionalized first through colonial legacy, then legitimized and formalized through the relations within the world capitalist system after colonialism. The other negative aspect of this form of development for Bunker is the co-opted internal state bureaucracies of peripheral countries that fail to protect from these conditions, partially as a matter of their own centralized bureaucratic structure. We have already documented that the world financial institutions further complicate this problem by setting up parallel systems of governance in peripheral states, using co-opted local elites as the World Bank did in Colombia from 1956 through 1963. As we noted in Chapter 2, much of the Bank's funds went to set up and run a parallel system of control separate from the Colombian government.

In sum, this ecological unequal exchange undermines nature for locals either to subsist upon or take to market for fair trade in commodities, thereby also undermining their very sustainability. Further, the core needs this relationship to continue its metabolism—its elite consumption of worldwide nature. Finally, although this unequal exchange is facilitated and orchestrated through core states and tribute states in the periphery, it is organized by state apparatuses, which connect through nodes, likely via commercial and bureaucratic hubs of urban areas.

Here we have explained the state's role in *industria*. It propagates a world capitalist political economy (which began first as a mercantile capitalist system and now is a market capitalist system) that privileges the profit motive for firms. Throughout this book, we have explained that the knowledge/power systems, or systems of power that protect and preclude certain knowledge and ways of being in the world, have been reduced to a single neoliberal ideology where there appears to be few or no competing alternatives. What we have not been told is that the only way this world capitalist system can operate is through the progressive enclosure of the commons, increasing penetration into the biosphere, and continued suppression of those who would dissent. As noted earlier, *industria* is a predatory and expansionary force that cannot survive without continued devouring of nature. It will not stop on its own or by those immediately benefiting from this system. It must be stopped from the outside.

It is now time for those outside of this state-corporate alliance to confront and change the character of states that have participated in the domination of nature and marginalized social groups that privilege a few at the expense of the planet. We contend that by understanding the state in this way through our green theoretical proposition, concerned people from around the globe can join their voices to force the state in each of their nations to change from its industrial role of domination and profiteering to one that is defined by a more protective role toward nature. At this point, no one group is powerful enough on their own to enact this change, this change can only come from a sort of transnational civil society that is conscious enough of the harm being done to the capacity of planet Earth to survive that they can stand up to modernity's demons. In *Lord of the Rings*, the ancient wizard Gandalf comes to a place in the story where a giant Balrog demon of fire descends upon him and his friends. As he protects his friends, he is resolute—"you shall not pass!" Unfortunately, our demons, like the hegemonic part of our states' character, are harder to see and much more secretive, but this chapter has worked to use theory as an illuminating sword to expose the monsters that threaten humanity and planet Earth. It is our time to stand up to the demons that drain away our critical life systems of planet Earth and say "You shall not pass!"

We have noted that civil society has grown within the confines of the state, but it need not remain chained to these boundaries. At one point in time, individuals and groups created the character of states, and we suggest that civil society today, particularly with new abilities to communicate with each other through social networks, is capable of forcing our individual states to move toward a more protective role for those few remaining wild places left. Seeing the state as a predatory and expansionary force that is designed to advance upon diverse ecologies and peoples to direct power to a "legacy elite" group of people socially related to those who were able to use their natural capital to build managerial and martial capacity for conquering and expanding out into other areas allows us to understand that the state is anything but a legitimate source of authority/violence, a stabilizing institution, or a source for sustainable environmental protection. But, it is still the only institution powerful enough to rein in the power of the real enemy, transnational corporations that exist in an abstract world without responsibility to any community.

Under the green theory of the state, we have shown there are coercive and ideological forces that have imposed certain ontological conditions, and that part of hegemony is when we volunteer to its demands often without even knowing we are doing it. That is because some conditions of public life come to be taken for granted, assumed to be "common sense," and are normalized—we do not question them. The dominant ideology that has put us into a zombie-like stupor is economism, a belief that unlimited economic growth through neoliberal market mechanisms is the answer to all of our problems. Yet, it is this very policy pursued by *industria* through state co-optation that is leading to the end of life on planet Earth as we know it. Even more insidious, this ideology of economic growth is often tied to nationalism. Nationalism can function as one of the most impenetrable barriers to a livable world when it operates as a form of deadly parochialism that blinds us to the legitimacy of others in the world. We may love our countries, and there is nothing wrong with that until it comes at the expense of others, who also have the right to love their country. Currently, the state system has instituted an exclusionary and very unequal world economic system that eats at the critical life systems of all of us, but our sense of nationalism often blinds us to the role our own states are playing in this nihilistic dance of death.

We can loosen the chains of nationalism by creating transnational associations with each other in a new embodiment of civil society. But civil society at any level is not an entity that emerges by some magical process. We have seen how the elite of European civil society emerged in concert with the state, joined by a common interest in the pursuit of power based on capitalist accumulation. New transnational associations must now emerge organized on noneconomic, moral grounds of the right of all life on planet Earth to exist and peoples to choose their own value systems and way of life. On a planetary level, this new civil society will only have agency to the extent that people within their individual spaces become conscious of the powerful forces that are destroying planet Earth, and question the sometimes jingoistic nationalism and the norms of economism. Civil society can only gain agency to stand and resist these forces through the conscious resistance of people and peoples who join together in their concern for planetary survival.

If the state is to remain, it must do so under a new social contract and new personality, which pursues the positive alternative, defends against appropriations of the commons, and controls firms. The state now is the only institution with enough power to challenge and keep firms under control, and unlike firms, the state retains some modicum or presentation of the public interest. It is the firm now that is on the rise, as it has been so fully empowered by the state system that the concentration of power organized first by the states is now being conferred to transnational corporations. The state still maintains capital in the public sphere, but we are at a critical moment where the priorities of the state must be changed to more fully recognize social and ecological values over economism, or we are going to live in a desolate landscape without forgiveness.

This new social contract can come in many forms, depending on the current structure of government apparatus and values of individual nations of peoples involved. What must be common to all is that the state must change its habits of privileging the profits of firms over everything else, and they must recognize that a global division of labor where the core consumes the periphery is not only unsustainable physically, but also unconscionable. A new social contract must uplift the values of diverse groups, and recognize the nonhuman nations that deserve to live as much as we do. We can never ignore the warnings of Gramsci of the coercive hegemony often practiced by states or the cautions from Bookchin regarding the inherent tyranny of state hierarchies, but that does not mean we resign ourselves to the belief that our states will not hear our voices, especially if they have some components of democracy.

So, how does civil society operate with or within the state? It performs a dialectical, evolutionary movement of give and take, between the state structure's limits and the demands from the people, as described by Anthony Giddens.[76] But, this dialectic only makes sense if there is a consciousness that *we are a people*—a public. And we contend that we are now at a point where we are a people who coexist on a single planet, and this is an important constraint on what one public should be able to do because there is no legitimate privileging of one person or group or public over others. No one group should be able to create a crisis at the population level. There are nations upon nations, none should be transformed into slaves or tools or resources because as in Plumwood's ecological self, virtue demands that we see these nations as "I's" as we see ourselves. Transnational civil society will only be able to confront the institutionalized power of the firm (privileged by the state system) if we gain the consciousness that we are a force and that we have the productive power to build holy associations that favor the planet and all of its children. When transnational groups of people develop a consciousness based on an ontology of joy rooted in a deep sense of interconnectedness with all life, we can be a global society that attains a dialectical force because we can see a common good in protecting the planet. A transnational public sphere can have agency when they know they are a public, and when they know they can challenge the state's, now exposed, disciplinary power, together.

But where can we look for leaders who have modeled strong community and recognition of nonhuman nations?

Indigenous Resistance to the State

It is not accidental that most examples of civil society's demanding a new relationship to the state apparatus are coming from peoples who live in the most intimate connection with the land, especially Indigenous communities. In the shadow of this predatory *industria* these subsistence economies, especially the Indigenous peoples, are refusing to give permission for ecological exchange with the core. We have seen some vivid examples of this refusal. In Seattle in 1999, we saw Indigenous and subsistence peoples stand side by side with unionized workers and young activists, dismissed by the co-opted corporate media as anarchists, to say "NO!" to the World Trade Organization. They were arguing that "another world is possible" as they tried to raise a world consciousness that current corporate-led globalization is destroying their way of life. Indigenous groups in Ecuador, in fact, mounted a broad-based social movement to demand explicit rights for nature. These rights are now institutionalized for the first time in the history of the state, in a national constitution, and anyone can bring suit on behalf of nature. In Ecuador, they have built one version of a new social contract.

The politics of modernity are, as they always have been, ever so much more than an issue of distribution. They are about identity and the availability to live life without being "administered" by Western hegemony. This is a more complete notion of justice and resistance in the face of this unequal ecological exchange, mounted by state-firm tyranny. The neo-Marxist brand of justice used so far only conceives of justice in terms of distribution, but what of "recognition" and participation? David Schlosberg convincingly demonstrates in his work that a rethinking of environmental justice is necessary to encompass more than just the distribution of goods and burdens, but to also include the reasons behind maldistribution, which are a lack of full recognition as part of the human community, and the ability to participate in decisions related to this distribution.[77] In the sense that the First and Third Worlds are both trying to work within and toward a global economy—the First World continues to appropriate value, and the Third World is often forced to work within international institutions and domestic politics that expect it to "catch up"—the world of Indigenous peoples may be one of the most important areas of resistance to this economism still around.[78] It is time to pass the talking stick to those who have felt the boot of this Western hegemony most vividly to gain the insight and courage that will be necessary for the path ahead.

Chapter Six
Earth Consciousness, Earth Action

> Join me in Earth Revolution.
> —*Ta'Kaiya Blaney (11-year-old girl of the Sliammon First Nation)*[1]

If we accept the proposition that the predatory world-system of *industria* is destructive and illegitimate, "what do we do now?" Thankfully, the contradictions of infinite economic growth on a finite Earth are being recognized in wider and wider circles. Ta'Kaiya, along with the rest of the *world* Indigenous movement, a *world* peasant movement, and a *world* social movement, suggest a planetary revolution toward *buen vivir*. This will take theory, warriors, and love to fight retrenchment, denial, and even desperate violence that may come with changes and challenges to the current world-system. In this chapter we will explain that we have traded away all our relations for materialism and an insatiable capitalist world-system. We will describe the benefits of an alternative ontological position (ways of living or sometimes "life-ways") in *buen vivir,* and pass the talking stick to Indigenous leaders who explain specific Indigenous life-ways. Then we will show that Indigenous leaders and their people have not gone quietly out of the international picture but have mounted an impressive world movement that is allied with the world peasant movement and the world social movement. All of these movements declare that "development" ideals and policies have failed, the current capitalist world-system is not only unsustainable; it is not worthy of promoting, and a transition to a more just world living in harmony with Mother Earth is the only alternative to total world collapse and chaos. We then conclude with some broad suggestions for moving toward this more democratic and peaceful alternative. However, first we will show that the current capitalist world-system is *already* changing.

Immanent Change of an Aging and Brittle Capitalist World-System

Immanuel Wallerstein argues that the loss of the Vietnam War and abandoning the gold standard for the US dollar indicated the first stages of decline. Ironically, abandoning the gold standard also initiated the neoliberal era.[2] Sociologists Smith and Wiest agree, and they reveal that the world capitalist system is facing increased crises that have grown over the last few decades, and these crises indicate a decline in the system itself. Challenges to the world-system are exemplified by widespread protests around the world, including those in the global North seen in riots in Europe and the Occupy Movement in the United States. These protests are reacting against corruption, obscene state protection of the most powerful sectors and people, crony capitalism, austerity programs, and economic ruin ultimately caused by "the policies of global and regional financial institutions, and thus in the long term the protests are likely to strengthen movements for changes in the larger world system."[3] Smith and Wiest's research indicates the world-system has probably planted the seeds of its own destruction. Legitimacy of mainstream institutions that prop up the world-system is being challenged at a number of levels. These researchers provide a long list of systemic contradictions: progress in human rights militates against the mainstream institutional rules of endless accumulation, where, for example, rising food and water insecurities demonstrate "incompatibilities between human rights and globalized markets"; the liberal promise of a thriving middle class is sacrificed to large capital gains and countered by historic inequities that neoliberalism promotes; ecological needs of the system to feed endless accumulation are untenable; and the failure of "development" in the global South is generating the call for alternatives—both nonviolent and violent projects, such as openly anti-Western international terrorism. Challenges to the system have prompted a more active and coercive set of policies from core northern states—such as aggressive incarceration of minorities, removal of civil liberties, and militarized borders and immigration policies. These actions have raised the cost of policing and military projects necessary to keep the core states in a hegemonic position internationally, which Smith and Wiest doubt can be maintained:

> Few would argue we are now witnessing a time of great crisis. The collapse of the global financial markets and increased uncertainty in the financial sector, the growing evidence of large-scale climate disruption and species extinctions, unstable and rising food and energy costs and large-scale inequality are coupled with scarcity of water and arable land and rising threats from international terrorism.[4]

They continue,

> These multiple and interrelated crises all can be seen to signal the physical and social limits of the existing world capitalist order.... [This] system's need for

constantly expanding markets and economic growth contends with the hard reality that we live on a single planet that is not growing, and that, although the productivity of workers can often be increased, there are physical limits to how much surplus value (profit) can be extracted from the planet and its people.[5]

Others agree, and these warnings are not from unstable extremists on street corners telling us to repent because the end is nigh. Rather, these are voices of reason, experience, and deliberation from many different perspectives. For example, Roberto De Vogli, a public health expert and senior lecturer at the College of London with a dual appointment at the University of Michigan, writes in his book, *Progress or Collapse: The Crises of Market Greed,*

> There are converging ecological crises looming on the horizon of modern civilization: climate change, peak oil, overpopulation, unsustainable consumption of water, fish and food. The world is on a collision course against the limits of the ecosystem. Modern societies are consuming, polluting and expanding as if there is no tomorrow. Indeed, there may not be one.[6]

Or take the work of John Casti, a scholar in the field of complexity who takes a bit of a different angle but comes to a similar conclusion about the future of the global system. Casti writes about the increasing world vulnerability in *X-Events: The Collapse of Everything*: "The infrastructure required to maintain a postindustrial lifestyle—power, water, food, communication, transportation, health care, defense, finance—are all so tightly intertwined that when one system sneezes, the others can get pneumonia."[7] In his view, collapse of the current system is inevitable. Modern human civilization has built such complex systems that they are vulnerable to an increasing array of random events that may not even have any historical precedent, such as the twilight of oil, global pandemic, or a collapse of the global food system consistent with our concerns in this book. Casti hopes people hear his warning in order to transition and prepare for this change so that it is not riddled with misery. However, if we do not we will suffer "total system meltdown, where in this case human civilization itself is the 'system.'"[8]

These are but a few examples of increasing but sober concern for the limited sustainability of the current world-system, yet leaders of the world-system have effectively ignored these warnings or at best treated them with ineffective piecemeal, incremental policies that attempt to, if temporarily, hold back crisis *to* the world-system not protect the Earth or its people *from the* world-system. Notably, we have more environmental laws now than ever before, but political economic conditions such as accountability in world trade, finance, and consumption are all but nonexistent, and structural environmental problems, and their attendant existential threats, continue to grow.

Now compare these academic perspectives with the following:

> In the absence of a true implementation of sustainable development, the world is now in a multiple ecological, economic and climatic crisis; including biodiversity loss, desertification, deglaciation, food, water, energy shortage, a worsening global economic recession, social instability and crisis of values. In this sense, we recognize that much remains to be done by international agreements to respond adequately to the rights and needs of Indigenous Peoples. The actual contributions and potentials of our peoples must be recognized by a true sustainable development for our communities that allows each one of us to Live Well.[9]

This is a remarkably consistent statement that comes from the Kari-Oca II Declaration from Indigenous leaders in 2012, as they met parallel to, but outside of, the twenty-year meeting for the reconsideration of the 1992 United Nations Conference on Environment and Development, the Rio+20 meeting in Brazil, with of course the addition that Indigenous peoples have some ideas about how to approach the current crises and reemphasize the need to live well.

Thus, while the timing is unpredictable, the world-system looks like it will change, perhaps in the next few decades, where two obvious alternatives are between a progressive politics from people of the land, or a militant assault on Western modernity. Clearly, we hope the former is stronger than the latter, but to encourage a nonviolent transition, Makere Stewart-Harawira calls us to our "Great Work" toward a livable planetary consciousness consistent with *buen vivir*.[10]

Our Great Work is needed not only to counter militarism but because *industria* is not a system that builds happiness. We pursue materialism to fill our sense of dissatisfaction, but this places us on a treadmill, and we are unfulfilled. But why? Why are we not happy?[11] We are not happy because we have traded the meaning of our lives away for Hummers, now ironically in disfavor; the endless aisles of Wal-stores; and other sprawling consumption. Worse, this unhappiness is promoted in the dominant discourses as "development" to the extent that the influential, modernist "stages of development" aspire to "mass consumption" as the end and finality of human progress.[12] We suggest, in concordance with other voices that are ignored daily in the world news and world conversations, that *this is not development, freedom, or progress,* and it is tearing the Earth and its inhabitants asunder. Indeed, the Indigenous People's Earth Charter (discussed below) notes on point 66, "The concept of development has meant the destruction of our lands. *We reject the current definition of development* as being useful to our peoples. Our cultures are not static and we keep our identity through a permanent re-creation of our life conditions; but all of this is obstructed in the name of so-called developments."[13]

We Have Traded Away All Our Relations

For the authors, a planetary consciousness is enmeshed in a way of being and seeing what is real—a universe that is "full." Following the Indigenous movement and La Via Campesina, we call this the ontology of *buen vivir*, or "living well." Living well comes from the joy of harmonious interconnection with all other Earth subjects—the two-legged, the four-legged, the winged, the conscious but normally ignored plants and rocks and the land. This ontology is inspired by prominent themes in the lifeways of Indigenous and peasant peoples and it seeks to hear their voices on subjects of living well on a changing Earth. Indigenous voices are not considered inert token artifacts for scrutiny, or icons of perfection (as in the so-called noble savage stereotype), but as human communities in the dynamic living world who have experienced and lived with great change—these histories lend wisdom and resilience and we need to listen. Human civilization can be inspired by Indigenous peoples' will and skill in surviving so long with meaning and purpose. Thus, this is a call to restructure our lives in smaller, more personal but transnational communities that fundamentally respect the other and attempt to reweave all our relations again.

What community are you responsible to? Our drive to create a hyper-individualistic globalism means that we are more likely to have relations like those proposed by the childish philosophy of Ayn Rand—where love is despised because it subjects our apparently brilliant individuality to limits.[14] British conservative and former prime minister Margaret Thatcher, ironically countering the founder of classical conservatism, Edmund Burke, once uttered the base logic of the neoliberal program when she said,

> I think we have gone through a period when too many *children* and people have been given to understand "I have a problem, it is the Government's job to cope with it!" or "I have a problem, I will go and get a grant to cope with it!" "I am homeless, the Government must house me!" and so they are casting their problems on society and who is society? There is no such thing! There are individual men and women and there are families and no government can do anything except through people and people look to themselves first.... If children have a problem, it is society that is at fault. **There is no such thing as society**. There is living tapestry of men and women and people and the beauty of that tapestry and the quality of our lives will depend upon how much each of us is prepared to take responsibility for ourselves and each of us prepared to turn round and help by our own efforts those who are unfortunate.[15]

First, this logic ignores the structural causes of misery, where any one person's opportunities are semi-dependent on history, education, transportation, skills, social connections, and our parents' resources, not to mention a plethora of other things outside of our control, but which are orchestrated at higher levels of society. These different opportunities are "life chances" that reflect what kind of income, healthcare, education, or retirement we have, and life chances are related to how much death and disease we encounter.[16] If death and disease are orchestrated by society itself, then society commits what Friedrich Engels called "social murder."[17] Research indicates that economic policy designed to "maximize the accumulation of profit while socializing the associated risks and costs" promotes social murder inasmuch as these policies produce "economic instability, unemployment, poverty, inequality, dangerous products, and infectious and chronic disease."[18]

Next, if we are only individuals and there is no society, then we have no responsibilities and limits are construed as threats to freedom. Taken to its logical conclusion, we might wonder if there is any problem cannibalizing our own neighbors. Obligations to neighborhoods, networks, or nightingales become burdens that unfairly snip the individual flower at its bud, even though that same flower has grown from decades of soil formation that are the contributions from so many others through the march of time. Under this logic, where obligations and limits are anathema to abstract freedom, anything goes, everything is for sale, and in a bitter turn of irony, nothing is free because all human purpose is incarcerated in a single authoritarian vision of the good. This vision is always hungry but never satisfied.

Meanwhile, as responsibilities and limits are shunned, the dominant logic of development expands power for core states, empires, and firms to capitalize on our inability to theorize and create meaningful, resistant collectives. Today, even after the concussions from the Great Recession of 2008, Rand's logic and its incumbent system is ironically *stronger* than it was prior to the crisis, despite being a cause of it. Author Gary Weiss writes, "Rand has experienced an extraordinary revival since the financial crisis, and nothing seems to be stopping her.... Yes, she was an extremist, but she matters because her extremism is no longer fringe."[19] While there is broad agreement that deregulation of the financial sector and related problems such as regulatory capture created the Great Recession, captains of the wreckage retrenched, denied, and attacked the implication that firms or the neoliberal system should itself be questioned.[20] For example, the largest bank in the United States, JPMorgan Chase, has come under scrutiny in civil suits (not criminal) for "widespread fraud" as it sold compromised goods, backed by bad investments (mortgages in overvalued property), but William K. Black, a law professor at the University of Missouri–Kansas City, notes that by October 2012, "They are not prosecuting any elites from Wall Street."[21] All the while, "Wall Street struck back against the very idea of reigning in its compensation [of corporate leaders who were obscenely compensated to march the system into crisis] and any curbs on its ability to freely transact business regardless of societal consequences,"²² apparently including social crisis spanning massive foreclosures, deep unemployment, and eventually deep austerity measures that put places such as Greece and Spain into civil chaos.

Ayn Rand's anarcho-capitalism can be seen in former US president George W. Bush, who indicated resistance to helping struggling homeowners facing evictions: "We got [sic] to be careful and mindful that any time the government intervenes in the market, it must do so with clear purpose and great care," but *at the same time* the US Federal Reserve bailed out JPMorgan Chase and its subsidiary Bear Stearns with $30 *billion*, which transferred about $12 *billion* to JPMorgan Chase shareholders.²³ Even Alan Greenspan, a devoted follower of Rand and grand architect of US free market absolutism, "admitted that he had put too much faith in the self-correcting power of free markets and had failed to anticipate the self-destructive power of wanton mortgage lending," which also allowed for "shadow banking," or management of money through unregulated venues like hedge funds or private equity firms.²⁴ In 2012, well after the Great Recession was initiated, the *New York Times* reported that "The so-called shadow banking system, blamed by some for aggravating the global financial crisis, grew to a new high of $67 trillion worldwide."²⁵ Here, the neoliberal system has generated crises at the population level, but little is being done to challenge the power of these elites, let alone challenge the ideology of Rand's radical capitalism that has shredded the social contract in the global North and suppressed it from forming in the global South. Roger Keil remarks on the overall process after so many contradictions to capitalism have now been issued:

> Brundtlands' World Commission on Environment and Development addressed the West's growing fears that development as it was known after 1945 was not delivering the goods: economic growth in an era of accelerated neocolonialism and imperialism was stalled as countries of the global South were mired in a debt crisis of unknown proportions, ecological problems abounded as populations still exploded (filling up the shantytowns of the larger cities), and violence was endemic inside and between developing nations as independence as well as the incipient dissolution of the Cold War blocs had set free the centrifugal dynamics of militarism, civil wars, and permanent revolutions.²⁶

We argue that to remake a sustainable and livable social sphere, every individual must accept the egotistical limit that they are not the only one who matters. When we realize that we, in industrial settler societies, cannot spread out to a new space and consume another people's space, we will find ourselves in better accord within sustainable limits and the law of ages.²⁷ These limits are filled with relations—and joy. Limiting ourselves through accountability and authentic relations with others in our world is a first law of virtue and the first step to broad worldwide joy. Limits come through accepted norms that are respected across nationalities as morally accepted mandates, and this type of norm also provides pathways for fulfillment of individuals. That is, norms develop based on how we expect each other to act in order to live together. The mainstream norms of *industria* are disastrous, promote misery and destruction, and it is time they changed to reflect *buen vivir*.

What Makes *Buen Vivir* Ontology Good or Joyful?

"Onto-" refers to being, and "ontology" refers to our notions central to being in the world, which condition our sense of purpose, reality, and the context for relationships we have as beings in the world. *Buen vivir* refers to building a life of fulfilling relationships with each other and the Earth. We have shown that the dominant sense of being in the world in the modern era has been economic, stripping away our relations to other beings in the world and to any possibility of a life of harmony and meaning along with these endless potential associations with others. If we listen to voices of Indigenous people, we can begin to hear that harmony with all our relations is central to being human and that we cannot arbitrarily remove the agency and importance of any being in the world without doing violence to our relations. If our relations are constitutive of our own being, then the violence we do to the Earth is violence we also do to ourselves and our purpose.

Joy is happiness regardless of circumstance. Joy is not found in trappings but in relationships. Naturally, the opposite of joy is also found in relationships, but ones that turn sour, where our relations have been neglected. The more peaceful and content we are in the multitudes of associations we have in the life and land and sea around us, the more joy we will live. If joy is found in relationships, stripping away the entities of the world with whom we might have a relationship impoverishes the social life we might lead for at least two reasons. One is that such stripping away allows for exploitation, and it is hard to maintain joy in the face of others' misery and suffering unless we are severed from our own compassion. Aldo Leopold forecast our situation in this sentiment:

> One of the penalties of an ecological education is that one lives alone in a world of wounds. Much of the damage inflicted on land is quite invisible to laymen. An ecologist must either harden his shell and make believe that the consequences of science are none of his business, or he must be the doctor who sees the marks of death in a community that believes itself well and does not want to be told otherwise.[28]

We see the marks of death all around us, most profoundly in the scars on the land and loss of nonhuman nations and the Sixth Great Extinction.[29] In order to live unaffected in *industria*, first we are forced to blunt or deny our sense of compassion to the suffering imposed on so many from the exhaustion and disposal of life support systems in the periphery, or we can face up to defending Mother Earth. Joy is lifted in compassion and withers outside the sun of virtue. Second, we abandon the relationships and entities around us, and lose the understanding that joy grows in companionship, fed in love. This is not the love of Eros, of passion, but of interconnectedness and a unity of existence that allows us to love even our antagonists. This is the love that Martin Luther King Jr. spoke of when he preached about love that existed even for those who were trying to beat him, suppress the African American community, and crush the larger civil rights movement. This unity does not mean that other's interests are the same, nor that we are "all one" unintelligible pot of melted identity. Blurring over the differences in the entities, people, and nonhumans impoverishes the universe of distinction. But, in our dominant world model, there is "no society," only alienated individuals so different they find no compelling reasons to associate meaningfully with others in the world, even with other humans. This is part of the rootstock of identitarian politics and the legitimacy that *industria* wields. Distinctions and features and traits that make us authentic individuals make us rich, but the richness is contextualized and built in communities on Earth. Every single distinction, every single individual, every single group is set within a larger, interacting context of space and ecology. Individuals find joy in the fullness of their relationships, and *industria*, in order to concentrate wealth and power to specific historical elites, has institutionalized the ontology of radical exclusion that separates associations between people and between the nonhuman communities. In this model, everything outside the dominant logic of the self becomes an essentialized Other available for commodification. This stripped-down narrow consciousness not only makes the residents of core *industria* unhappy, it has brought and continues to bring misery upon the peripheral nodes in *industria* every day for the past 500 years. Ultimately, it is destined for biological failure and global collapse because there is only so much of the world that can be consumed before the whole is swallowed and the Sixth Great Extinction is complete.[30] *Buen vivir* is "living well" because it allows for earth others like fish and trees and mountains as well as human others space and legitimacy to their own life, and the sense of the good life is embodied in our virtue and relationships with earth others, not on gross accumulation, *and* it is a sustainable way to live together on Earth.

Passing the Talking Stick: Indigenous Lifeways— Respect for All Our Relations

> Despite being devalued, marginalized, disenfranchised and frequently submerged throughout the history of Western imperialism, traditional

> Indigenous knowledge forms have a profound contribution to make towards an alternative *ontology* for a just global order.[31]

In passing the talking stick, we see that Indigenous people around the world have been working to be heard.

> Often it is written that indigenous peoples have a spiritual relationship to their territories. This is not a myth.... The Elders say that the land is our mother. From the land and the territories, the knowledge of the relationship to all things is learned. —Sharon Venne/Old Woman Bear (Cree)[32]

Given this relationship, and the fact that Indigenous people witnessed both the dispossession and destruction of so much of their land, Indigenous leaders call upon us to join them and quite literally stop *industria* in its tracks.

In this section we will explore specific ontologies and specific lifeways of Indigenous peoples, and attempt to pass the talking stick to Indigenous leaders. However, one caveat should be kept in mind—in societies with talking stick traditions, voices are spoken not written. Communities have been small enough to hear individual voices, but now we live in a world too large for this and we must write these voices down and publish what is said—and this is not entirely consistent with the notion of the talking stick. Nonetheless, writing is what we are pragmatically confined to if we are to communicate *buen vivir* to a broad enough audience that Earth revolution is possible, and even Indigenous leaders from oral traditions publish their words on the Internet to be heard.

That said, here we will attempt to highlight what the ontological position of *buen vivir* is and how consistently it is believed around the world by an incredibly diverse group of Indigenous peoples. These are ontologies because they indicate the purpose of life and the key laws for peace and fulfillment on Earth in our lives together. Several Indigenous ontologies are examined, attempting to use authentic Indigenous voices as a way to hear what specific tribal peoples believe are the key elements to a sustainable global consciousness.

Our call for a new ontology fits what Val Plumwood calls the "ecological self":

> The truly social self is the mutual self; the social self salutes the social other as another *self*, a centre of subjectivity like mine but a different one, one which imposes limits on mine, and incorporates this salutation into the concept of "I".... Similarly the ecological self recognizes the earth other as a centre of agency or intentionality having its origin and place like mine in the community of the earth, but as a different centre of agency, which limits mine.[33]

The following examples of *buen vivir* show that there is opportunity for industrians to listen, learn, and integrate some lessons from Indigenous ontologies into the lives we lead in our own spaces to eventually overturn *industria* altogether. Mayan scholar Victor Montejo writes,

> It is my hope that we may learn to value and recognize indigenous knowledge expressed by Elders when they explain Mayan beliefs and spirituality with emphasis on values: such as the respect that links humans, nature, and the supernatural world. By knowing and valuing these ways of life, we will understand the suffering that indigenous people have endured for the past five centuries.[34]

To better understand Indigenous politics, it is important to understand one of the most basic social structures used by Indigenous peoples to persist: this structure is tribalism.[35] Tribalism is usually not constructed through "blood relations" but is rather constructed through "kinship relations." And kinship is broadly construed, where tribes regularly adopt members of other tribes, as well as strangers.[36] This begs the question for industrians everywhere—"who are your kin?" This is not a question of who your blood relatives are—but who are your kin? How do we recognize our kinship with others? How are we responsible to our kin? *Industria* has significantly blocked the ability to have meaningful responsibilities to extensive kinship relations, and this is one aspect of the poverty and the vulnerability of *industria*.

The tribe is a network of people who have expressed responsibility and who rely upon each other. While tribes had their own ceremonies and cultural arrangements, Sharon O'Brien notes in the classic textbook in *American Indian Tribal Governments*, that "despite the differences in structure, however, traditional tribal governments shared certain values, ideas of leadership, and styles of decision making."[37] Tribal structure provided many Indigenous peoples egalitarian, decentralized (though not universally), and highly democratic governance based on rule of the people toward harmony with the supernatural, ecological, and human worlds:

> Traditional Indian cultures made little distinction between the political and the religious worlds. Political wisdom was synonymous with religious power. All political actions were undertaken with spiritual guidance and oriented toward spiritual as well as political fulfillment. Indian religious and political values were based on the belief that a spiritual force lived within every natural being. *The primary goal of religion and politics was to achieve harmony between all elements—the land, plant and animal life, and the human community. Human beings were not considered superior to and above nature but rather were thought to be connected to and part of nature.* People did not own the land and resources; instead, individuals had a responsibility toward all aspects of life.[38]

Traditional tribal government was based on institutional norms of responsibility and harmony where community responsibility came before individual rights or privilege. Power was based on consent from the community, and was usually given to elders for their wisdom, experience, or skill. Vine Deloria notes that tribal structure is inextricable from Indigenous social structure. Tribes provide localized governance and institutions that observe and reinforce the rules of living with the land so that one life does not wantonly consume others, or place others in unnecessary jeopardy. Tribes also provide immediate accountability for life practices, where tribal institutions reflect what has worked for the tribe in the past—usually a very deep and long history of continuity. Tribes, therefore, provide a way to keep survival and continuity at the forefront based on close observation and experiential knowledge.[39]

In a review of Native Americans and the environment, David Rich Lewis writes that the relationships maintained in Indigenous societies of the Americas were grounded and rich with dense ecological associations in a holistic world where people were, at least partially, constituted by their relations with the rest of existence:

> Native Americans have long had an immediate relationship with their physical environments. At contact most lived in relatively small units close to the earth, cognizant of its rhythms and resources. They *defined* themselves by the land, by the sacred places that bounded and shaped their world. They recognized a unity in their physical and spiritual universes, the union of natural and supernatural. *Their origin cycles, oral traditions, and cosmologies connected them with all animate and inanimate beings, past and present.*[40]

Indigenous peoples managed this world's bounty and diversity based on years of accumulated wisdom—the trial and error of previous generations in specific places. They acknowledged the Earth's power and the reciprocal obligation between hunter and hunted. They acted to appease spirits who endowed the world. Native peoples celebrated the Earth's annual rebirth and offered thanks for her first fruits. They ritually prepared the animals they killed, the agricultural fields they tended, and the vegetable and mineral materials they processed. They used song and ritual speech to modify their world while physically altering that landscape with fire and water, brain and brawn. They did not passively adapt, but responded in diverse ways as individuals and groups to refashion environments to meet their cultural and material desires.

This is set in contrast to the European states that organized structures of hierarchy to mobilize and command people and ecology to gain power for an elite world minority described throughout this book.

To examine specific Indigenous values, we will start with Maori perspectives, with the caveat that these brief portraits are oversimplifications of ancient perspectives that could not fully be reproduced here but are themes articulated mostly by Indigenous peoples about themselves.[41]

Maori

> Most indigenous peoples believe that the fundamental starting point is a strong sense of unity with the environment. Arising from the close and enduring relationship with defined territories, land, and the natural world, and exemplified by the pattern of Maori adaptation to Aotearoa (New Zealand).
>
> —*Mason Durie (Maori and Pasifika)*[42]

Durie believes that a primary characteristic of Indigenous peoples everywhere is an "enduring relationship between populations, their territories, and the natural environment," which provides the cultural knowledge, languages, practices, and ethics of that group, including his own, the Maori. A common understanding through Indigenous knowledge is the inalienability and inseparability of physical and spiritual worlds. Stewart-Harawira notes that these worlds are sewn together through *Te Aho Tapu*, or the "sacred thread."[43] The physical and spiritual are drawn together in complex notions of emergence, existence, and return. As all of existence is conditioned together in these universal relationships, all of existence is seen as having life force, or *mauri-ora*, which is

> A concentration of life, like the centre of an energy source ... for everything to move and live in accordance with the conditions and limits of existence. —Sir James Henare (respected Maori elder)[44]

> This concept of *mauri* as the unique living force that is present in all kingdoms of existence extends to inanimate as well as animate objects and, indeed, to concepts and forms of knowledge. Within the natural world, each individual rock and stone, each individual animal and plant, as well as every body of land and water, is recognized as having its own unique life-force. —Makere Stewart-Harawira (Maori-Waitaha/Scottish)[45]

In addition, each element with a life force is believed to have a supernatural guardian, which is "an expression of the need to acknowledge and protect the life force present in all aspects of the natural world."[46] Durie notes that this connection with *mauri* means that it "spirals outwards seeking to establish communication with higher levels of organization and to find meaning by sharing a sense of common origin."[47]

This sense of being is so full and profound that it is difficult to imagine a more holistic way of being and way of knowing. It tells us that all people need to defend the land, the water, the animals, and the plants as an expression of life on Earth. Also, if we fail to be guardians, our lives will lose the ability to maintain harmony with all of existence. Such an ontological position is incompatible with much of the Western philosophical tradition of alienation and separation of human and nonhuman worlds of being, and the Indigenous ontology presented here is expressly incompatible with a life of imperialism, wanton disposal and tyranny of others through the past 500 years of Western domination and ecological despoiling.

> The same technological revolution that has saved or transformed millions of lives through medical advances and technological miracles has, through blind obedience to the vagaries of market capitalism brought humanity to the abyss of despair over its future.... Collectively, humankind stands poised on the brink, facing the greatest moment of choice of all time—a plunge into the abyss of our own destruction or a great leap forward in consciousness. *There is no viable third way here....* The options are to undergo a radical change in consciousness and evolve new ways of being, or face certain extinction of many species and a drastic diminishment of others ... at the core of Indigenous ontologies lies a deep understanding of the meaning and practice of interconnectivity and of spirit that is vital in this stage of the transformational journey of humankind.[48]

Mayan peoples

The key ways of knowing in Mayan culture surround specific places, the communal structure of life, and stories that relate how Mayan culture has survived. The Mayan cosmology is rooted in a multitude of associations:

> When we Mayans say that we respect nature, we are sincere, because we live what we say; we feel a unity with other living creatures on earth. It is not only that we are "close to nature" but that we recognize the value of life, and we respect others. The Mayan cosmology and world view are centered on communal practices in which all elements that promote life—cosmic elements (e.g., sun, wind), humans and the environment are interrelated. Our world view also informs the politics of how we must act and react when this communalism is threatened by outside forces. —Victor Montejo (Mayan)[49]

For this reason, Montejo writes that when Mayans who observe the ways of the Elders want to change the world around them, they must ask permission. For example, if a Mayan campesino (peasant) wants to cut a tree to plant corn (the symbol of humanity), "He asked permission to the 'Giver of Life' to cut down the trees for the cornfield."[50] "Appreciation and respect" then frame these actions, *because* there is a unity with other living creatures and recognition of others. The campesino *still* needs and creates the cornfield, but he is not morally permitted to take more than is sufficient—enough for his family's and village's needs—because others need that space and the lives taken to make the cornfield *matter*. Unlimited tree cutting wantonly disposes of this life without recognition of the loss, and it undermines the "elements that promote life" so that "those who cut trees for pleasure shorten their lives."[51]

We might remember that Montejo's tradition has had the sober experience of catastrophic collapse of the classic Mayan complex society—which existed at least from 2000 BCE—beginning its rapid descent about 790–890 CE.[52] During this time, the Mayan population across the Yucatan in Mexico and Guatemala dropped from 3 million to about 450,000. Theories for the collapse of the classic Maya include deforestation and over-intensity of agricultural practices resulting from a population grown too large for the resources available in the Mayan agricultural lowlands. Thus, we might listen closely to the words that Montejo and the modern Maya have about lessons learned from living on a limited and changing Earth. Related to the Mayans are the Southwestern tribes in the United States. Some Pueblo people recognize a direct kinship with Mayan peoples.

Pueblo peoples of the US Southwest

The Pueblo tribes have lived for more than 10,000 years as descendents of the Mogollan and Puebloan peoples (referred to as Anasazi by some, but which means enemy ancestor in Navajo). The Puebloan peoples may have abandoned their complex architecture of cave dwellings due to climate changes, which brought on drought and populations that could not be fed and then collapsed. The Pueblo tribes believed that "there is a spiritual force within all of nature. Nature and God are one. Humankind's task is to maintain a harmonious relationship with nature."[53] Pueblo tribes also ruled by consensus with leaders providing counsel, not commands.

The Hopi of current Northeastern Arizona are well known for their spiritual governance that springs from the ancient dictates of creation stories and laws. Wall and Masayesva describe some of these dictates and the nature of life on fragile arid mesas for about a thousand years, referring to the current era as the "Fourth World" that humans emerged into after the three prior worlds were destroyed by avarice, greed, and neglect:

After their Emergence into the Fourth World, the clans that would one day comprise the Hopi people approached the Guardian Spirit, Masaw, in the region that is now northwest Arizona and asked his permission to settle there. Masaw recognized that the clan people's former life, which they knew was not bringing them happiness, had been given over to ambition, greed, and social competition. He looked into their hearts and saw that these qualities remained, and so he had his doubts that the people could follow his way. "Whether you can stay here is up to you," he told them.

Masaw warned the clan people that the life he had to offer them was very different from what they had before. To show them that life, Masaw gave the people a planting stick, a bag of seeds, and a gourd of water. He handed them a small ear of blue corn and told them, "Here is my life and my spirit. This is what I have to give you."[54]

There is a distinction between the one true Hopi, Masaw, and the people who follow his way. Masaw is the true embodiment of a Hopi; the people who follow his way are merely Hopi *Senom,* or People of the Hopi. Following common tradition, however, members of the Hopi tribe discussed in this book will be referred to as "Hopi."

To be Hopi is to embrace peace and cooperation, to care for the Earth and all of its inhabitants, to live within the sacred balance. It is a life of reverence shared by all the good people of Earth, all those in tune with their world. This manner of living lies beneath the complexities of *wimi,* or specialized knowledge, which can provide stability and wisdom but when misused can also foster division and strife.

Deeper still in the lives of traditional Hopi people lies the way of Masaw, a way of humility and simplicity, of forging a sacred bond between themselves and the land that sustains them. Masaw's way is embodied in corn. At the time of the Emergence, Masaw offered the clan people a manner of living that would not be easy. Dry farming in the high desert of northern Arizona, relying only on precipitation and runoff water, requires an almost miraculous level of faith and is sustained by hard work, prayer, and an attitude of deep humility. Following the way of Masaw, the Hopi people have tended to their corn for nearly a millennium, and the corn has kept them whole.

For traditional Hopis corn is the central bond. Its essence, physically, spiritually, and symbolically, pervades their existence. For the people of the mesas corn is sustenance, ceremonial object, prayer offering, symbol, and sentient being unto itself. Corn is the Mother in the truest sense—the people take in the corn and the corn becomes their flesh, as mother's milk becomes the flesh of the child. Corn is also regarded as the child, as when the wife of a farmer tends to the seeds and newly received harvest, blessing and ritually washing the corn, talking and singing to the seeds and ears. The connection between the people and the corn is pervasive and deeply sacred. In a remarkable symbiosis between the physical and the spiritual, the Hopi people sustain the corn and the corn sustains Hopi culture.[55]

Central to life in Hopiland is corn and water. Kachinas, who dwell in the mountains commonly referred to as the San Francisco Peaks, bring this water if they have lived according to law. Corn represents the very Earth, Mother. Here corn is another "I" that is essential for the self-reference of being Hopi, as described by Wall and Masayesva, and the corn cannot be taken for granted any more than the land, or any more than another Hopi or non-Hopi person. Here the corn is a "sentient being unto itself," but corn also symbiotically forms Hopi culture, which shows how all our relations co-create who we are; if our relations are sick or worse denied and cut-off, we cannot be well. For the Hopi people, in order for them to be well, they believe they specifically need to tend to the relations of corn and all that makes up corn—land, water, spirit, love.

Yakama (formerly spelled Yakima)

Similarly, the Yakama ancestors lived in the Columbia River Valley of the current United States for more than 14,000 years. The Yakamas were a hunting and gathering people organized around the seasons. O'Brien writes, "The Yakimas recognized the beauty and sanctity of all nature and felt a deep spirituality for and emotional attachment to the land. Gratitude for nature's bounty and the desire to give thanks and protect it formed the basis of their life."[56] Salmon was a principle staple, and the Yakama believed that "they were not fish but godlike people who lived beneath the sea."[57] The fish are not a simple resource, they are other people—in other words, they are full agents in the world with their own purpose and functions that must be respected and thanked as recognition for the laws and limits to the human community and its consumption of others.

Lakota

Likewise, the Lakota people of the Northern Plains of the United States, particularly the Badlands and Black Hills, have a strong affiliation with the land. Lakota leaders—Sitting Bull, Red Cloud, and Crazy Horse along with Cheyenne leader Roman Nose—all successfully defended this land in the Powder River War of 1864 forcing the US government to make a treaty in 1868. However, within six years, the treaty was violated as Lt. Col. George Armstrong Custer was ordered to search the Black Hills for rumored gold. Within months the Lakota were told they had to sell this land or it would be taken, but the Lakota refused. The land was invaded, and in 1876, Crazy Horse and Sitting Bull defeated US troops in the Battle of Rosebud and then again in the well-known Battle of Little Bighorn, where Custer and his entire company were killed. Unable to defeat the Lakota, mass slaughter of the bison, a staple for the tribes, was encouraged, leading to mass starvation. The virtual extinction of the plains bison pressed the Lakota into the illegal agreement of 1876. Sitting Bull never surrendered, though, noting,

> The Lakota ... loved the Earth and all things of the Earth, the attachment growing with age. The old people came literally to love the soil and they sat or reclined on the ground with a feeling of being close to a mothering power ... the old people liked to remove their moccasins and walk with bare feet on the sacred earth. Their tipis were built upon the earth and their altars were made of earth. The birds that flew in the air came to rest upon the earth and it was the final abiding place of all things that lived and grew. The soil was soothing, strengthening, cleansing, and healing. What law have I broken? Is it wrong for me to love my own? Is it wicked for me because my skin is red? Because I am a Sioux; because I was born where my father lived; because I would die for my people and my country?[58]

Fourteen years after the 1876 agreement, Sitting Bull returned from Canada, where he had escaped. In December 1890, he was arrested while camped at the Standing Rock Reservation. In the meantime, a band led by Big Foot was surrounded and marched to Wounded Knee on the Pine Ridge Reservation.

This incident deserves some further explanation due to its continued importance in the world Indigenous movement. While weapons from the band were collected, the medicine man Yellow Bird began to chant, and others followed with their death song. At this point a shot was fired from either the US troops or one of Big Foot's Ghost Dancers. The troops unloaded their weapons on eighty-four men, forty-four women, and eighteen children who were massacred at Wounded Knee. Some were hunted down and found dead more than a mile from the scene.[59] Since that time, the Lakota were forced to abandon their nomadic hunting life for one of settled agriculture. It was at this same location in 1973 that the American Indian Movement (AIM), including the late Russell Means, came to occupy the reservation against the ruling regime of Richard (Dickie) Wilson who was openly against Oglala traditionalists of his tribe. The federal government, allied with Wilson, faced down the occupation with assault rifles and full military force, but the occupation lasted for months. At the end, the Wilson regime continued. During this time, a paramilitary group working for Wilson, called the GOON Squad, along with the US Bureau of Indian Affairs, "unleashed a reign of terror on the tribal chairman's adversaries. More than seventy of Wilson's opponents died violently."⁶⁰ In 1975, AIM returned to protect the embattled residents of the reservation who were, quite literally, being hunted down in the name of modernization of the tribe and the valorization of this centralized tribal government. During this incident, two FBI agents were murdered. Three men were charged for these deaths; two were acquitted and the third, Leonard Peltier, continues to serve a double life sentence despite what former United States attorney general, Ramsay Clark, says was fabricated evidence.⁶¹ It was this bald violence toward traditional tribal values, even when it was abetted by the Lakota's own council leader, that garnered international support from traditional cultures in favor of AIM's efforts (see below). AIM indeed lost that particular battle, and the corrupt structure of government and the antagonisms against traditional culture still inhabit Pine Ridge today, but not without a grim toll.

O'Brien notes that by 1989 the unemployment rate was around 80 percent and 65 percent of families were living on less than $3,000 a year.⁶² In 2012, Pulitzer Prize–winning journalist Chris Hedges and acclaimed illustrator Joe Sacco visited Pine Ridge to find that these statistics had not changed in over a decade. Pine Ridge still suffers from 80 percent unemployment and the same rate of alcoholism, and the average man lives to forty-eight years old alongside a tsunami of violence—"Rape and indiscriminant violence are the legacies of white conquest."⁶³ None of these circumstances would have come to pass if the US government had not used industrial force to extirpate the bison. The bison, which at one point had a population between 25 and 30 million, were extirpated in a punctuated slaughter at the end of the nineteenth century, and the skins were shipped to Europe to supply new industrial processes that provided a demand for the leather while the carcasses of the bison were left to rot on the plains.⁶⁴ But, to the Lakota, the buffalo were not just food, hides, or tools for control; the buffalo represented another nation of people. The close connection between the Lakotas and *pte oyate* kin (the buffalo nation) is evident in the Lakota story of the coming of the White Buffalo Calf Woman to the Sans Arcs band. The story begins with a time of famine: The Sans Arcs, moving westward, are unable to find buffalo. Two young men, sent out to find game, encounter a beautiful young woman, who explains that she has been sent by the Buffalo tribe with a message for their people. One of the two lusts after the woman and is destroyed. The other returns to the camp, tells the people what he has seen, and they prepare to welcome the visitor. The next day the woman appears at sunrise carrying a pipe. After being welcomed by the chief, she takes the pipe and explains that Wakan Tanka (God)⁶⁵ has smiled upon everyone present, as all belong to one family. She, in fact, is their sister. The woman then explains that she represents the Buffalo tribe and that the pipe she carries is a gift from them to the people. After instructing the women, children, and men of their duties and obligations, she lights the pipe and offers it to the earth and the four directions.

She puffs on the pipe, passes it to the chief, and then leaves. As the people watch her depart, the woman suddenly becomes a white buffalo calf.[66]

Further,

> Lakotas and Nakotas regarded buffalo as spiritual beings that had originated within the earth and were relatives of humans. Like the many other wakan beings of the world (including animals, birds, insects, stones, and thunder), buffalo were capable of appearing and speaking to people in visions or dreams.... The White Buffalo Calf Woman reminds the male hunters that the necessities of life come from the "earth below, the sky above, and the four winds." Yet receiving these necessities is not automatic. "Whenever you do anything wrong against these elements," she instructs, "they will always take some revenge upon you." Therefore, "You should *reverence* them. Offer sacrifices through this pipe. When you are in need of buffalo meat, smoke this pipe and ask for what you need and it shall be granted you." Scarcity, then, was explained as a consequence of people offending "the elements." Abundance could be secured through ritual. Not simply a mechanical activity, ritual efficacy required adherence to moral values.[67]

Chief Arvol Looking Horse, the 19th Generation Keeper of the Original Sacred White Buffalo Calf Pipe of the Lakota, Dakota, and Nakota Nation of the Sioux, perhaps the foremost expert on the issue, notes,

> Nineteen generations ago the beautiful spirit we now refer to as Pte-san Win-yan (White Buffalo Calf Woman) brought the Sacred C'anupa to our People. She taught the People the Seven Sacred Rites and how to walk on Mother Earth in a sacred manner. She said, "Only the good shall see the Pipe ... the bad shall not see it or touch it."[68]

Not only are bison related to humans and form their own tribe or nation, but in order to live well and have sufficient food and health, this tribe needs to be respected, honored, and *revered*. If this does not occur, then humans will not only be trespassing on other peoples (bison and other wankan), humans will suffer. George Tinker tells us these sophistications may be one of the most important observations for sustaining a livable world:

> Some sense of what is at stake is apparent in a Lakota phrase that may be illustrative. *Mitakouye oyasin* can be translated as a prayer "for all my relations." As such it is inclusive not only of immediate family or even extended family, but of the whole tribe or nation; of all the nations of two-leggeds in the world; and particularly of all the nations other than two-leggeds—the four-leggeds, the wingeds, and the living-moving things. It is this interrelatedness that best captures what might symbolize for Indian peoples what Euro-Americans would call creation. More to the point, it is this understanding of interrelatedness, of balance and mutual

respect among the different species of the world, that characterizes what we might call Indian people's greatest gift to Euro-Americans and to the Euro-American understanding of creation at this time of ecological crisis in the world.[69]

It is clear that we could go on and on, looking to the words of Native scholars describing the ways in which Indigenous ontology includes all our relations, sees the varied actors in the world (canyons, rivers, salmon, bison, humans) as literally interrelated and constitutive, not figuratively and not metaphorically. Much of the idea of traditional ecological knowledge (TEK) has evolved around the specifics of Indigenous knowledge about the world around them. However, it is not just knowledge—as "ontology" indicates—it is a lifeway. Anishnabe scholar and writer Deborah McGregor reminds us that TEK is more than knowledge; it is about relationships and "instructions from time immemorial and on generations of careful observation within an ecosystem" and is a *practice*.[70] And, since the land, the knowledge, and the people are an integrated whole that make up TEK—Indigenous people are needed as active communities and co-creators in a changing world to *do* TEK.[71] *Industria* cannot just squeeze Indigenous peoples for compartmentalized effective information as they do in biopiracy. Truly consulting Indigenous peoples for active decision-making and collaboration means that the current world-system cannot continue with the status quo.

From the South and Central American Mayan to the plains tribes of North America to the Northwestern Yakama peoples to the Maori of the Pacific, we see that there is a common but differentiated theme that is articulated and lived in different ways. Nonetheless, this theme exists—Indigenous ontologies have at their foundation a living creative universe that is filled with others who are like other "I's," and this full universe provides limits and expectations of respect upon humanity—Indigenous or non-Indigenous people alike. We can decide how to practice our varied specific values and vital cultural stories, but we must all do this in accordance with the universal mandate to allow others to do the same. Currently, *industria requires* violating these universal mandates, and on these and other grounds, is a death pact many do not realize they have consented to. Our plea to you, dear industrian, is to remove your consent from this pact and begin reestablishing our relations so that Earth and its life may heal. You will not be alone—Indigenous leaders and peoples have initiated and sustained a working anti-imperial *worldwide* movement that can lend important ideas and leadership if only the talking stick is taken seriously and we are allowed to hear them. In the next chapter, we will explain the character of this movement.

Chapter Seven
A World Indigenous Movement

"We Are the Watchers, We Are Witnesses"[1]

In order for a transnational, cosmopolitan justice and a world community to emerge and live in joy, we believe that several things must change. First, we are in desperate need of a planetary consciousness that values all our relations. This means that the hour is here when we must reject the ontology of alienation to adopt a more peaceful and just way of living. This does not mean that we all must live the same way or that we must all be the same—in the previous chapter, we heard from a diverse array of native beliefs that have many ways of living, varied governmental traditions, varied economic systems, and different religious traditions. But all Indigenous societies have an answer to institutions, or social rules, that inscribe a respect for life on Earth. This does not mean that all tribal people are virtuous—that would be racist as well as inaccurate because tribal people are people like everyone else—but what it does mean is that *Indigenous people answer to a sense of order that industrians have abandoned*. A sense of order based on respect for all life on planet Earth is an essential first step in identifying a world community.

Second, through the mainstream politics of alienation and never-ending industrian goals of endless expansion and capital, we are destroying the Earth or at least desolating it substantially. This is not a slogan, it is not a triviality, and it is not a lonely sentimental echo from environmental conspirators. There are parts of the Earth we are quite literally tearing asunder, and this message comes from many different voices around the world. For example, after the

1992 United Nations Conference on the Environment (Earth Summit) in Rio (which Indigenous peoples could not officially negotiate in; see below), Hopi elder Thomas Banyanca cautioned that "human beings are destroying Mother Earth in pursuit of money and greed."[2]

A vivid example of this destruction is presented in a research article in the journal *Science*.[3] The article stated that "dead zones" have increased from a handful in the 1960s to more than 400 in 2008 and may turn out to be a key issue in future dynamics of life in the ocean. Dead zones result from the hypoxia (lack of dissolved oxygen [DO]) that occurs when there is a large growth in planktonic algae spurred by nutrient enrichment from nitrogen and phosphorous fertilizer pollution. This fuels microbial respiration as the algae die and float to the bottom, depleting DO in the water. Dead zones are called this because there is not enough DO to sustain life in these hypoxic or anoxic areas, resulting in various levels of marine mortality or migration. Dead zones are driven by the "explosive growth" in "industrial produced nitrogen fertilizer that began in the 1940s."[4] This was precisely the beginning of the Green Revolution, as argued in Chapter 4, where the United States, through the US Department of State and the Rockefeller Foundation, exported the industrialization of agriculture first to Mexico then to other areas in the periphery starting in the 1940s. This was not done out of benevolence but to make food mechanized and develop agricultural science "as a tool to construct hegemony" as it inserted this science of control into food, culture, and power.[5] If this is not destruction of an Earth system, it is hard to imagine what is—and thus it is not some empty bumper-sticker slogan to say that we are destroying the Earth because through multiple layers of impact from human-driven global environmental change, humans are reducing biodiversity on the entire planet in addition to altering the life support systems in the oceans, freshwater, on land, and in the sky.[6]

The ontological understandings of Indigenous peoples described in the prior section are not about past or extinct cultures, but of living contemporary cultures that have long genealogies that can only come from effective adaptation and wise living. Bradley Reed Howard writes that we are now experiencing "a resurgence of activism among Indigenous peoples, energetically asserting their international rights not only as individual human beings but as self-determining peoples, unique and independent cultures."[7] We will now explore some of the cross-national Indigenous rights movements specifically articulated against the vanities of modern capitalist accumulation of ecological beings and spaces and the colonization of other peoples.[8]

Continued international networks and groups that make up the global Indigenous movement are acting on a planetary consciousness of defending the Earth and the self-determination of Indigenous peoples to live free from imperialism. Additionally, we have already cited and explained other subsistence-based movements that often include Indigenous groups, represented by groups such as the Third World Network. Another important example of the world Indigenous movement is Survival International, an international nongovernmental organization that describes itself as "a movement for tribal peoples" spanning more than eighty countries. It has won the Right Livelihood Award, otherwise known as the alternative peace prize, for its work advancing tribal rights, and is clearly a key organization in the world Indigenous movement.

One of the earlier Indigenous groups is the International Indian Treaty Council (IITC), created by ninety-seven Indian Tribes and Nations on the land of the Standing Rock Sioux tribe in the United States. This is documented in the Declaration of Continuing Independence by the First International Indian Treaty Council in June 1974. Moved by the siege at Wounded Knee in 1973 where the Lakota nation and the AIM made a stand for Indian rights and sovereignty, this group formed to create an international prosecution of the US government for genocide and to call attention to the loss of Indigenous freedom and land by colonial powers around the Western Hemisphere. They implored the international community to listen and hear Indigenous voices that have been suppressed just as Mother Earth was being used and commandeered for gross exploitation:

> Sovereign people of varying cultures have the absolute right to live in harmony with Mother Earth so long as they do not infringe upon this same right of other peoples. The denial of this right to any sovereign people, such as the Native American Indian Nations, must be challenged by *truth* and *action*. World concern must focus on all colonial governments to the end that sovereign people everywhere shall live as they choose; in peace with dignity and freedom.[9]

Today, the IITC continues to intervene on the behalf of Indigenous peoples for cultural and human rights related to Indigenous tribal sovereignty, including the right to live in harmony with Mother Earth, as noted above. In passing the talking stick here, we can listen to people with ancient lineage to specific places on the Earth that want to be able to live according to the "right of any sovereign people" but have been kept from doing so. We hear that all people have the "absolute right" to live in harmony with Mother Earth "so long as they do not infringe upon this same right of other peoples." *Industria* quite clearly can operate *only* if it violates both of these tenets, and ideologically must convince its subjects that Indigenous peoples should not enjoy self-determination and therefore have no right to live as they want and need to. IITC has been arguing with a depth of voices from around the world for all people to support Indigenous peoples and their struggles, and to end colonial practices. However, settler governments are unlikely to stop and listen unless people within the mainstream political populace (industrians) ally themselves with groups such as the IITC and argue from within that whole political structures must be rearranged. Ultimately, these ancient voices echo a truth that some groups of people cannot rightly consume the ecological space of others. Consequently, we are forced to be self-sufficient in peacefully occupied spaces through fair trade and exchange that do not annihilate shadow ecologies.

Another example of the global Indigenous movement that links demands for revolutionary changes in lifeways with land, people, and all our relations is the Kari-Oca Declaration. The declaration was produced during the Rio Earth Summit in 1992. Indigenous peoples, as colonized political groups that have been subject to the state system, were not allowed to participate as sovereign nations in the Rio summit, so Indigenous peoples from around the world convened their own meeting and came up with their calls for action that were much stronger than those coming from the state system. The following is the totality of the declaration:

> We, the indigenous Peoples, walk to the future in the footprints of our ancestors. From the smallest to the largest living being, from the four directions, from the air, the land and the mountains. The creator has placed us, the indigenous peoples, upon our Mother the Earth.
>
> The footprints of our ancestors are permanently etched upon the lands of our peoples.
>
> We, the indigenous peoples, maintain our inherent rights to self-determination. We have always had the right to decide our own forms of government, to use our own laws, to raise and educate our children, to our own cultural identity without interference.
>
> We continue to maintain our rights as peoples despite centuries of deprivation, assimilation and genocide.
>
> We maintain our inalienable rights to our lands and territories, to all our resources—above and below—and to our waters. We assert our ongoing responsibility to pass these onto the future generations.
>
> We cannot be removed from our lands. We, the indigenous peoples are connected by the circle of life to our lands and environments.
>
> We, the indigenous peoples, walk to the future in the footprints of our ancestors.[10]

At the Kari-Oca meeting in 1992, Indigenous leaders produced the Indigenous People's Earth Charter with 109 points aimed at cultural sovereignty and rights and ecological respect for Mother Earth.

The first point is, "We demand the right to life," under the section of "Lands and Territories." These first points refer to a spiritual interrelatedness to the land that is given by the Creator in which there is a balance and mutual respect among different species of the world that must be protected.

31. Indigenous Peoples were placed upon our Mother, the Earth, by the Creator. We belong to the land. We cannot be separated from our lands and territories.
32. Our territories are living totalities in permanent vital relation between human beings and nature. Their possession produced the development of our culture. Our territorial property should be inalienable, unceasable and not denied title. Legal, economic and technical backup are needed to guarantee this.[11]

Again, in 2002, parallel to the state-based World Summit on Sustainable Development (WSSD) meeting in Johannesburg, Indigenous leaders from the around the world issued the Kimberley Declaration from the lands of the Khoi-San in current Southern Africa with unified voices:

> We continue to meet in the spirit of unity inspired by the Khoi-San people and their hospitality. *We reaffirm our mutual solidarity as indigenous Peoples of the world in our struggle for social and environmental justice.*[12]

In this forum they again indicate,

> We the indigenous Peoples of the World assembled here reaffirm the Kari-Oca Declaration and the indigenous Peoples' Earth Charter. We again reaffirm our previous declarations on human and environmental sustainability. Since 1992 the ecosystems of the earth have been compounding in change. *We are in crisis. We are in an accelerating spiral of climate change that will not abide unsustainable greed. Today we reaffirm our relationship to Mother Earth and our responsibility to coming generations to uphold peace, equity and justice.*[13]

The leaders remind us that these declarations are not new but affirm prior declarations, committees, affirmations, negotiations, and other international law, including the Declaration on the Rights of Indigenous Peoples adopted by the United Nations General Assembly in 2007;[14] the Charter of the International Alliance of Indigenous and Tribal Peoples of the Tropical Forests; the Mataatua Declaration; the Santa Cruz Declaration on Intellectual Property; the Leticia Declaration of Indigenous Peoples and Other Forest Dependent Peoples on the Sustainable Use and Management of All Types of Forests; the Charter of Indigenous Peoples of the Arctic and the Far East Siberia; the Bali Indigenous Peoples Political Declaration; and the Declaration of the Indigenous Peoples of Eastern Africa in the Regional WSSD Preparatory Meeting.[15]

Finally, in 2012, as noted in Chapter 1, Indigenous peoples came together for the Rio+20 conference (the United Nations Conference on Sustainable Development), which met as a follow-up to the 1992 Rio Earth Summit. Again, Indigenous leaders were not allowed to be official participants like nation-states, but they met alongside this meeting prior to the start of Rio+20. More than 500 Indigenous people signed the Kari-Oca II Declaration, but beyond that, they blessed it in ceremony to make the words sacred, and then brought it to the state-led meeting demanding to be recognized as essential actors, committed to *buen vivir,* that are opposed to simply greening capitalism to make it appear more sustainable (the disappointing goals of Rio+20 were to promote a greening of economic trade and production):

> We, Indigenous Peoples from all regions of the world have defended our Mother Earth from the aggression of unsustainable development and the overexploitation of our natural resources by mining, logging, mega-dams, exploration and extraction of petroleum. Our forests suffer from the production of agro-fuels, bio-mass, plantations and other impositions of false solutions to climate change and unsustainable, damaging development.... Mother Earth is the source of life which needs to be protected, not a resource to be exploited and commodified as a "natural capital." We have our place and our responsibilities within Creation's sacred order. We feel the sustaining joy as things occur in harmony with the Earth and with all life that it creates and sustains. We feel the pain of disharmony when we witness the dishonor of the natural order of Creation and the continued economic colonization and degradation of Mother Earth and all life upon her. Until Indigenous Peoples rights are observed and respected, sustainable development and the eradication of poverty will not be achieved.[16]

It is not surprising that the urgent, consistent, and persistent claims of the world Indigenous movement were ignored at Rio+20 because the United Nations meeting had no intention of really challenging the system that had brought the crises to begin with, and if they considered the voices from Kari-Oca I or Kari-Oca II, this reflexivity would have been necessary.

Again and again, international Indigenous leaders attempt to rein in the exploitation of modern *industrial* colonial and neocolonial efforts that undermine all our relations, and remind us that limits to our own industrian consumption and use of the planet Earth are essential to justice, sustainability, and peace. Without these limits, we forget who our relations are, we forget the land, we forget who we are as inhabitants of a world filled with others, and we jeopardize our ability to continue to live in this place. It is time that we pass the talking stick and listen to these voices. It is not too late for industrians to join their voices with the world Indigenous movement and demand that this world-system no longer suppress dissent and that, for example, the Declaration on the Rights of Indigenous Peoples be observed by the full international society of states, that Indigenous nations be provided with space as sovereign nations in the United Nations, and that all authentic—not illegal or coerced—treaties with Indigenous and non-Indigenous peoples be received as international law. We might also be wise to listen to the Indigenous People's Earth Charter; below are some examples:

95. Indigenous wisdom must be recognized and encouraged.
96. The traditional knowledge of herbs and plants must be protected and passed on to future generations.
97. Traditions cannot be separated from land, territory or science.
98. Traditional knowledge has enabled indigenous Peoples to survive.[17]

As part of a practice in the ontology of joy, we entreat everyone who sees all our relations as definitive for life on Earth to practice reciprocity with the rest of existence, to respect the spatiality of our lives, and defend the commons against insatiable capital in *industria*. These changes will end the consumption vital to the metabolism of *industria* and it will collapse, allowing new human potentials to be set free.

Ramparts of Consciousness and Commons

It is time to have a broken heart. It is time to allow our sorrow to awaken the wolves of our consciousness, to awaken our warrior spirit of guardianship. To the extent that we are willing to defend and guard against exploitation and to tell the truth in a violent world, we are warriors. Warriors are not people who go out and fight to kill for wanton destruction—that is more aligned with the current system of neoliberalism and militarism—but are rather those individuals who are willing to stand against the powerful to defend the weak, the sick, the exploited, the enslaved, children, the Earth, and the future generations. As we participate in *industria*, we are told that "nature" is different from "civilization" and humanity, but when the wolf awakens in us we know the wolf is kin, and that the wolf, like the Earth, looks on from the brush waiting for the moment to reveal itself. Separation of people from land and the rest of existence is a lie slipped into the veins of industrians like heroin, immobilizing us with the drugs of mass consumption. It seems, then, that to sell off all our relations, we cut out the vision and power to resist the predatory world powers, but to awaken our relations and to recognize the nations of animals and plants around us in a clarion to counterhegemony. George Tinker offers this first rampart for modern—*contemporary*—liberation and sustainability:

> My contention will be that attention to these two important spiritual aspects [reciprocity and spatiality] of Indian cultures and what I am calling Indian theology can become radically transformative for the Euro-American system of values and structures of social behavior.[18]

Tinker explains that reciprocity is the bedrock of balance. We see the ideas of reciprocity throughout the Indigenous ontological concerns first noted in this chapter, and the global Indigenous movement described above. Reciprocity is the idea that humans are part of a cosmic whole, where we are neither separate nor above the rest of existence, and that each human action has an effect in this existence even if we do not understand it. Tinker says, "Knowing that every action has its unique effect has always meant that there had to be some sort of built-in compensation for human actions, some reciprocity."[19] Thus, because existence is full of other "I's"—we must respect other agents and nations, and any effect on this world must be balanced by human compensation. Politically, this imparts a limit on human actions, and, if observed, would make *industria* impossible on its own because *industria* is concerned with rapacious consumption for its own sake, without compensation to the rest of existence. Indeed, *industria* cannot effectively face elements of existence like rivers or rhinoceroses as partners in a balanced world at all. It simply wants to devour the rivers and rhinos, which it largely has, without accountability to the nation of rhinos, the individual rhinos, or the rest of existence for their extirpation. If we had to pay in real terms—not money but prayer, gifts, self-deprivation and sacrifice, *time,* and ceremony—for our incursions and actions in the world, we would be forced to account to the idea of balance and we could *begin* to repair the global environmental changes we have embarked upon, as much as is possible. Certainly, when we are discussing 500 years of incursion and unbalanced impact on the world from European states, empires, and *industria,* there remains a great deal to compensate for and doing so will not be done overnight, but it must begin now. As Tinker warns, "as far as I know there is no ceremony for clear-cutting an entire forest."[20]

Such paradigmatic shifts only begin when we forcefully erect the ramparts of reciprocity—a singular barricade against *industria* that will be effective and just. To build such a barricade against the further growth of *industria,* we must become warriors of imperfect virtue—people who will tell the truth, pay compensation for our impacts, pray for insight and permission, give time and deliberation to our actions, and demand that structures of core power in the world give in to balance and harmony themselves. If this is accomplished, the core powers will no longer be "core" but assemblies of people who account for their actions among themselves and the rest of existence. This change in transnational organization would be a way to collapse *industria* without collapsing actual communities through misery. Without reciprocity, none of our efforts toward sustainability are worthwhile.

Tinker calls attention to the other theme apparent in this and the last chapter's account of what Indigenous people have been saying when they hold the talking stick, which is that Indigenous people and cultures are committed and constituted in particular spaces. Space is not abstracted but concrete grounding for the creation of culture, rules, and identity; space is the landscape, the sky and water, the weather and ecology of a specific area that is situated in the cups of larger areas connected to the whole Earth.

> Each nation has some understanding that it was placed into a relationship with a particular territory by spiritual forces outside of itself and thus has an enduring responsibility for that territory, just as the earth, especially the earth in that particular place, has a filial responsibility toward the people who live there. Likewise, the two-leggeds in that place also have a spatially related responsibility toward all others who share that place with them, including animals, birds, plants, rocks, rivers, mountains, and the like.[21]

Thus, colonial removal from ancestral lands is not only a political-economic dispossession for Indigenous peoples; it has been one of deep spiritual and existential crisis. Similarly, because *industria* has disposed of its kinship ties to the Earth and its inhabitants and landscapes, the actors in *industria* have felt little need to guard these spaces or think of the land as anything more than a resource, for itself only.

Spatiality is a crucial notion, and if it is successfully defended in contemporary and future politics, *industria* will fall. Spatiality requires that we are responsible for our impacts in a space; this is impossible if we do not even know where our food comes from. In a world where production and consumption are fully alienated through the funnels of industrian infrastructure and commodity chains of economic globalization, there is no respect for spatiality. Defending the land and landscapes will also have the effect of turning our attention to those spaces that we subsist on. In particular, under the current global economic system, commodities are produced, transported, and consumed in distant places that hide the shadows of consumption—their impact on the health and well-being of people and nonhumans in the various places that these shadows are cast.[22] This distanced trading system is also incredibly wasteful in energy used for transport. Our commons are crucial life support systems for us and the rest of existence. They are embedded in a tapestry that cannot be undone without our own undoing. Through distant production of industrian consumption, we consume other people's life support; in liberating the world from such a system, we must then produce and consume within an integrated space.

We are now in a position to learn brutal lessons from the Mexican campesinos (peasants), Indian subsistence farmers, and indigenistas that were the first harmed by the Green Revolution. Many lost their life support through commodification and industrialization of their crops, often losing their land to enclosures of private multinationals through debt or encroachment from the state, and then enclosures of their seeds. Here we can learn from and ally with Indigenous communities around the world in their struggle for autonomy and self-determination through solidarity in the commons, defending it from encroachment and enclosure by *industria*. In this way, the dispossession of Indigenous lands was one of the first steps in the establishment of *industria*; now, this expansionary force continues to move outward and into the farther reaches, for example, genetic structures, working to enclose all or most remaining communal ecology. From these lessons it is critical that we defend and reclaim commons from enclosure and commodification—these are the areas communities depend on for life support systems in air, water, and land.

Commons are communal life supports and must never be destroyed. There are two fundamental ways in which commons are spoiled. One is where a community allows for over-use because rules in the community are not accountable to the community's dependence or the ecological system needs. This is the so-called tragedy of the commons. The second is when they are enclosed from the community and only open for individuals to use and dispose of as they want. When commons become private, the accountability is centered on the individual, who then can preserve or exploit the land, water, or air as s/he wants. Use of the system is viewed in the short term, at the exception of the community, including the nonhuman community. The latter is the larger danger, as the current economistic system continues to segregate land from people.

The commons are socially constructed in the sense that we socially frame and consent to what is removed from communal use. One way to fight against the destructive forces we have described is to work on the level of words to reclaim what has been commodified. Campaigns and social movements, particularly by the youth, are critical for reframing our ecological expectations. Economism is so triumphant, so hegemonic, that we expect the elements of existence to be cordoned off and "developed." Social movements in particular are important for challenging the legitimacy of hegemonic power, and associations of people within and across *industria* have the opportunity to work to specifically undermine its credibility.

We find commons in the air, on the land, and in all water; life on Earth in biodiversity; the structure of life on Earth in genetic foundations; and food. Life support systems have been usurped by industrian firms, which are consuming our requisite needs and all our relations at the same time. In defending the key life supports in regional areas, through transnational social networks and movements, we can develop a transnational civil society. Such an alliance with the global Indigenous movements and others can help us focus on the interconnected needs of all ecologies on the planet Earth, allowing us to insure that the conditions for life on Earth may be preserved in whole. It is time to erect the ramparts of public life—our life together—and defend the critical commons that we all need to live against their commodification for state-corporate disposal. These ramparts are in our words, our minds, and through our bodies. We will need to disrupt *industria* by creative nonviolent command of what has wrongly been stolen from the commons and enclosed for profit all the while continuing to do critical theory and practicing our own situated knowledge about how we can live self-sufficiently in one space without consuming other spaces. We must raise our voices to insist that our states reverse their alignment with the transnational corporations that are trying to privatize all of our commons for their own profit, and when our voices are not enough, we can use our bodies to occupy space under threat as the peasants of the Himalayas did in the Chipko movement.[23]

However, we are in danger of not noticing. Today, we bottle water from public sources, but care little for the protection of the public water (defend the tap!). We think that if we can buy our way out of water pollution at the tap, we will be safe—despite the fact that bottled water is no safer than tap water, and is often bottled with it anyway. This is what Andrew Szasz calls "reverse quarantines" seen in the organic food movement, the move to purify private air, and especially in bottled water.[24] All these efforts lead to a false sense of security, allowing us to think that we can protect ourselves through purchasing *public* safety. Then, alarmingly, we lose our public sense of defending critical ecological sources and lives.

A 500-year-old movement to enclose the commons is coming to its fulfillment. Currently, corporate power defended by state hierarchical and ideological power is moving to enclose some of the most important commons. Seeds and genetic heritage is one such move, where the second phase of the Green Revolution is using genetic modification and patenting to claim the very structure of life as private stock, private commodity. The structure of life is therefore for sale, and we are told it is good for us, it will solve world hunger just as the first phase was supposed to, and that it is the right of corporations to have their due, which are our lives.

The commons are not only essential for life on Earth—stable climate through the atmospheric commons, marine commons, freshwater for ecosystem functions, not to mention human thirst—commons serve the poor first. That is, things that can be privatized are then enclosed from use by all; those with the affluence to purchase the rights to bottled water then can claim that water from anyone else. Water in the commons does not require affluence for access, but rather operates on civic and social institutions—rules of access and use specific to communities. Poverty does not keep one from using mangrove commons in the still traditional parts of Southeast Asia, but in contemporary Southeast Asia these areas are being enclosed for shrimp farming. The shrimp go exclusively to the markets that pay for them, mostly Japan and the United States; meanwhile, the poor lose their life support system where they found charcoal and fish, among other necessities. Now where do they turn? Where do the fish go for nurseries? What of the lost rare and precious mangroves? What of the animals that lived there? In the end, these shrimp ponds are so polluted that they become useless even for shrimp farming in five to ten years, but the loss of the mangrove forests are mostly permanent and the devastation to the peasantry is debilitating and disempowering as they are forced to abandon cash-poor, but self-sufficient living for wage labor in a system where they have little training and little value to the market apparatus.[25] The mangroves, as public commons and as our relations, is an example of what needs to be rescued from the maw of *industria* as it scoops up whole peoples and ecologies for temporary satisfaction for a more demanding and persistent gluttonous savagery.

Now What?

To defend the commons, local people can resist and fight commodification, but given the power of globalism and *industria,* local people will need transnational support and mobilization as local powers pressure the local subversives to end the disruption. *Industria* has built electronic and physical infrastructure that can be used against *industria*, via instantaneous organizing through the Internet, mobilizing through mass populist communications, and through creative culture jamming of consumerism. In this way, we can use the rhetoric of *industria*'s "commitment to democracy" as juxtaposition to actual practice, and can continue to think and theorize about how to form a self-conscious civil society that will contest the power of the state and firm, and form a new social contract.

If the vision of a peaceful and more sustainable world, full of all our relations, appeals to you as a more fulfilling life, it is time for you to come out, remove your consent to the old social contract, and join the battle against those looking to sell life on earth as they would a tool or toy. Many of the tools to organize a new transnational civil society have been articulated by the World Social Forum, which calls for a protection of the commons; for capital and corporations to release their death-grip on politics and commerce; for a canceling of debt held by poor countries; and defense of biological and cultural diversity.[26]

If you agree with our assessment—that Mother Earth and the human prospect are being eaten by a system put in place by a concentrated but illegitimate power, and that the ontological world proposed by the world Indigenous movement is worth listening to—then we find ourselves in the familiar territory that ends so many discussions on sustainability—"so, what now?"

This discussion usually ends up with the fear that "if we dismantle industrial society we must all go back to lifestyles of the Stone Age." That is not what we are proposing, and it is not what the world Indigenous movement has proposed. What to do going forward and living together on the planet means forging a synthesis of what has been and what can be. Markets—some markets are egalitarian and allow for opportunities, some markets concentrate wealth and power. Political institutions—some institutions such as the nation-state funnel power to a minority over a large mass of citizens, others such as school districts or soil conservation districts or village councils can be more democratic. Technologies—some technologies truly reduce important hazards and difficulties without raising terrible social and environmental costs and may lower social and environmental costs, such as engineering designs offered by Paul Polak[27] and groups such as Engineers without Borders/Ingenieurs sans Frontieres, who make tools that break poverty cycles and allow for life-saving opportunities in cleaner water and sanitation, while providing groups of people true opportunities in local markets. The link between political, economic, and technological freedom is scale, and the relationship of scale to political power. In the 1970s, anarchist Ivan Illich noted in *Energy and Equity*[28] that as energy and power increase in scale, personal power is enslaved by a political elite who are able to divert benefits in the name of efficiency while creating vast inequality and pollution. Discussions at the World Bank or in any national leader's office are limited and managed from above without consultation of those who are affected because industrial society favors large authoritarian, technical, political, and economic scales that homogenize differences, such as varieties of rice and the rice cultures responsible for cultivating them, into monocultures of language, knowledge, and ways of life. The Leviathan is not sustainable or just in any of these categories.

We need a movement of people across borders to break up Leviathans and bring technology, markets, and politics to smaller scales. However, as Illich warned, "Liberation which comes cheap to the poor will cost the rich dear," and this means that power can be defended from radical forms at these high scales and, no doubt, they will be.[29] But, the contradictions of *industria* mean that it *must* change, even if the rich want to avoid it, because there is only so much oil and so much soil. Some of the outcomes of this liberation appear fairly clear. There has to be a worldwide move away from the Green Revolution, and more sustainable agroecological practices for growing diverse foods for diverse peoples. This means that resuscitating and recovering diverse knowledge that lives through distinct languages and culture are clear imperatives.[30] We need energy systems that do not impoverish the future or create energy castes, but that provide for movement and basic needs. We need political institutions that do not sacrifice citizens for consumers; and, political institutions must allow for discourse where a talking stick makes sense, and all those concerned are consulted and really heard about decisions that affect them and the future. All of these specific things are needed for any future human prospect, but none of this means that the future will resemble the past, like the Stone Age.

There are a few tactics that may work:

1. Ideological rebellion. The first job is to free our minds and the discussion. Ideologies most dangerous are those not spoken of, but left to the undercurrents of our minds, public assumptions, and the everyday instruments of the state. The first act is to "decolonize our mind" as an act of rebellion, and publicly question the ideology of growth. Growth, in public discourse, is taken as an axiom that should be drawn out and discursively disrupted and challenged through thoughtful consideration and comparison to alternatives. Alternatives come from other ways of thinking, and for that we need to consult different groups and people.
2. Radical consultation. The way discussions occur at large-scale organizations such as the World Trade Organization, International Centre for Settlement of Investment Disputes, the World Bank, or in national legislatures and parliaments, is not inclusive. Central to our book's argument is that injustice breathes and lives off of exclusion, and to be more inclusive is to seriously consider the values and interests of others. Though these large-scale institutions are not going away very soon, we must demand that Indigenous nations be included in these discussions as full participants. For example, at the United Nations, tribes and Indigenous groups, should they choose, should be recognized as independent sovereign nations and full participants alongside the nation-states. As we discuss and consult, we acknowledge that our perspectives are not universally valid but we might agree on core ontological values. It is under these urgent conditions that First Nation's (Attawapiskat) Chief Theresa Spence began a liquid-only hunger strike as a way to demand an audience with the Canadian leadership threatening Indigenous sovereignty and land to defend Mother Earth, with the demand that it is time to be "Idle No More."[31] This initiated a sub-movement to the larger world Indigenous movement, Idle No More, which shows signs of internationalizing to other countries.[32]
3. Make it real. We can convince others who do not agree with us by talking *virtue*. If we ask, "What does it mean to be human?" or, "What does it mean to be good?" we open the possibilities of connecting with our dialogue through a shared sense of humanity—and this move is already happening. See the fantastic work of the Spring Creek Declaration, which starts, "A truly adaptive civilization will align its ethics with the ways of the Earth. A civilization that ignores the deep constraints of its world will find itself in exactly the situation we face now, on the threshold of making the planet inhospitable to humankind and other species. The questions of our time are thus: What is our best current understanding of the nature of the world? What does that understanding tell us about how we might

create a concordance between ecological and moral principles, and thus imagine an ethic that is of, rather than against, the Earth?" It concludes, "If hope fails us, the moral abdication of despair is not an alternative. Beyond hope we can inhabit the wide moral ground of personal integrity, matching our actions to our moral convictions. Through conscientious decisions, we can refuse to be made into instruments of destruction. We can make our lives and our communities into works of art that express our deepest values."[33]
4. Most of us see and believe there is a greater good that is better than simple profit. Even the CEOs of large companies might be tempted to agree to this over dinner, and that might be the entré to changing the boardroom and the stateroom.
5. Remake the rules. When we hear each other out and identify what vision of the good we might be able to agree on, we set the stage for connecting expectations to these core Earth and human values, and these rules will take on a life of their own. If we cannot agree on new rules at these small scales, we should continue discussing, using the talking stick.

The goal of *buen vivir* is to begin to live together with all our relations well, on our home planet Earth, and to do this, Indigenous voices have concrete suggestions:

> This inseparable relationship between humans and the Earth, inherent to Indigenous Peoples, must be respected for the sake of our future generations and all of humanity. We urge all humanity to join with us in transforming the social structures, institutions and power relations that underpin our deprivation, oppression and exploitation. Imperialist globalization exploits all that sustains life and damages the Earth. We need to fundamentally reorient production and consumption based on human needs rather than for the boundless accumulation of profit for a few. Society must take collective control of productive resources to meet the needs of sustainable social development and avoid overproduction, overconsumption and overexploitation of people and nature which are inevitable under the prevailing monopoly capitalist system. We must focus on sustainable communities based on indigenous knowledge, not on capitalist development.[34]

We will conclude with a messenger of peace, Chief Arvol Looking Horse:

> We are the watchers. We are witnesses. We see what has gone before. We see what happens now, at this dangerous moment in human history. We see what's going to happen—what will surely happen—unless we come together: we—the Peoples of all Nations—to restore peace and harmony and balance to the Earth, our Mother....
> This new millennium will usher in an age of harmony—or it will bring the end of life as we know it. Starvation, war and toxic waste have been the hallmark

of the Great Myth of Progress and Development that ruled the last millennium. To us, as caretakers of the heart of Mother Earth, falls the responsibility of turning back the powers of destruction. We have come to a time and place of great urgency. The fate of future generations rests in our hands....

Our Prophecies tell us that we are at the Crossroads. We face chaos, disaster, and endless tears from our relatives' eyes, or we can unite spiritually in peace and harmony. It's time to bring the Message of the urgent need for Peace, of creating an energy shift throughout the world....

As Keeper of the Sacred C'anupa Bundle, I ask for your prayers for Global Healing. Our Mother Earth is suffering. Her wonderful gifts [of] the water, the trees, the air are being abused. Her children the two-legged, the four-legged, those that swim, crawl and fly are being annihilated. We see such atrocities occurring everywhere....

We must comprehend in each of our hearts and minds the two ways we human beings are free to follow, as we choose: the good way, the spiritual way—or the un-natural way, the material way. It's our personal choice, our personal decision—each of ours and all of ours....

You, yourself, are the one who must decide. You alone can choose. Whatever you decide is what you'll be—good or bad. You cannot escape the consequences of your own decision. On your decision—yes, on your own personal decision—depends the fate of the World.[35]

Notes

Preface

1. Laura E. Donaldson, "Writing the Talking Stick: Alphabetic Literacy as Colonial Technology and Postcolonial Appropriation," *American Indian Quarterly* 22, no. 1/2 (1998).
2. Throughout this book, we capitalize "Indigenous," following several scholars, such as Oland, Hart, and Frink (2012), who treat Indigenous as a reference of identity and therefore as a proper noun.
3. Johan Rockstrom et al., "A Safe Operating Space for Humanity," *Nature* 461, no. 7263 (2009); Johan Rockstrom et al., "Planetary Boundaries: Exploring the Safe Operating Space for Humanity," *Ecology and Society* 14, no. 2 (2009); Mathis Wackernagel et al., "Tracking the Ecological Overshoot of the Human Economy," *Proceedings of the National Academy of Science (PNAS)* 99, no. 14 (2002); Peter M. Vitousek et al., "Human Appropriation of the Products of Photosynthesis," *Bioscience* 36, no. 6 (1986); Peter M. Vitousek et al., "Human Domination of Earth's Ecosystems," *Science* 277 (1997); F. Stuart Chapin III et al., "Consequences of Changing Biodiversity," *Nature* 405, no. 6783 (2000).
4. Andrew Dobson, "Listening: The New Democratic Deficit," *Political Studies*, early online edition (2012).

Chapter One

1. Vine Deloria and Daniel R. Wildcat, *Power and Place: Indian Education in America* (Golden, CO: Fulcrum Resources, 2001), 63.
2. S. Levitus et al., "World Ocean Heat Content and Thermosteric Sea Level Change (0–2000 m), 1955–2010," *Geophysical Research Letters* 39, no. 10 (2012): L10603. See also S. Levitus et al., "Global Ocean Heat Content 1955–2008 in Light of Recently Revealed Instrumentation Problems," *Geophysical Research Letters* 36, no. 7 (2009).
3. Levitus et al., "World Ocean Heat (0–2000 m)," L10603.
4. M. Winton, "Does the Arctic Sea Ice Have a Tipping Point?" *Geophysical Research Letters* 33 (2006).

5. Quoted in Fred Pearce, "Arctic Sea Ice May Have Passed Crucial Tipping Point," *New Scientist*, no. 2858 (2012), www.newscientist.com/article/dn21626-arctic-sea-ice-may-have-passed-crucial-tipping-point.html.

6. Jeremy B. C. Jackson, "Ecological Extinction and Evolution in the Brave New Ocean," *Proceedings of the National Academy of Sciences* 105, suppl. 1 (2008).

7. Reg A. Watson et al., "Global Marine Yield Halved as Fishing Intensity Redoubles," *Fish and Fisheries*, early version (2012), http://onlinelibrary.wiley.com/doi/10.1111/j.1467-2979.2012.00483.x/abstract.

8. "Hottest Year on Record in the U.S. Might Still Be 2012, Despite Cooler October," *Huffington Post*, November 9, 2012, www.huffingtonpost.com2012/11/9hottest-year-on-record-us-america-2012_n_2.

9. "U.S. on Track for Warmest Year on Record, and Second Most Extreme," *ThinkProgress*, November 12, 2012, http://thinkprogress.org/climate/2012/11/12/1177671/us-on-track-for warmest-year-on-record.

10. "Hottest Year on Record."

11. James Hansen, Makiko Sato, and Reto Ruedy, "Perception of Climate Change," *Proceedings of the National Academy of Sciences* 109, no. 37 (2012): E2415–E2423.

12. See Baiqing Xu et al., "Black Soot and the Survival of Tibetan Glaciers," *Proceedings of the National Academy of Sciences*, early ed. (2009).

13. See M. L. Parry et al., eds., *Climate Change 2007: Impacts, Adaptation and Vulnerability. Contribution of Working Group II to the Fourth Assessment Report of the Intergovernmental Panel on Climate Change* (Cambridge: Cambridge University Press, 2007).

14. Carolyn Kormann, "Retreat of Andean Glaciers Foretells of Global Water Woes," (2009), e360.yale.edu. See also Claire L. Parkinson, "Earth's Cryosphere: Current State and Recent Changes," *Annual Review of Environment and Resources* 31, no. 1 (2006).

15. Krista Tippett, "Joanna Macy: A Wild Love for the World," National Public Radio (2010), http://being.publicradio.org/programs/2010/wild-love-for-world/transcript.shtml. See also Joanna Macy, *World as Lover, World as Self* (San Francisco: Paralax Press, 1991).

16. Peter M. Haas, "The Political Economy of Ecology: Prospects for Transforming the World Economy at Rio + 20," *Global Policy* 3, no. 1 (2012): 96.

17. Raymond Clémençon, "Welcome to the Anthropocene: Rio+20 and the Meaning of Sustainable Development," *Journal of Environment and Development* 21, no. 3 (September 1, 2012): 311–338.

18. Stephan Leahy, "Indigenous Message to Rio+20: Leave Everything beneath Mother Earth," *National Geographic News Watch*, June 19, 2012, http://newswatch.nationalgeographic.com/2012/06/19/indigenous-message-to-rio20-leave-everything-beneath-mother-earth/.

19. Ibid.

20. World People's Conference on Climate Change and the Rights of Mother Earth, "The People's Agreement," April 22, 2011, Cochabamba, Bolivia.

21. "United States Data," *World Bank* (2012), http://data.worldbank.org/country/united-states.

22. Ting Wei et al., "Developed and Developing World Responsibilities for Historical Climate Change and CO_2 Mitigation," *Proceedings of the National Academy of Sciences* 109, no. 32 (August 7, 2012): 12911–12915.

23. G. P. Peters, G. Marland, C. Le Quéré, T. Boden, J. G. Canadell, and M. R. Raupach, "Rapid Growth in CO_2 Emissions after the 2008–2009 Global Financial Crisis," *Nature Climate Change* 2 (2011).

24. Peter J. Jacques and Jessica Racine Jacques, "Monocropping Cultures into Ruin: The Loss of Food Varieties and Cultural Diversity," *Sustainability* 4, no. 10 (2012): 2970–2997.

25. L. Maffi, "Bio-Cultural Diversity for Endogenous Development: Lessons from Research, Policy and On-the-Ground Experiences," *Endogenous Development and Biocultural Diversity* (2007), 56 (emphasis added).

26. W. J. Sutherland, "Parallel Extinction Risk and Global Distribution of Languages and Species," *Nature* 423, no. 6937 (2003).

27. Kent H. Redford and J. Peter Brosius, "Diversity and Homogenization in the Endgame," *Global Environmental Change* 16, no. 4 (2006): 317; J. P. Brosius and S. L. Hitchner, "Cultural Diversity and Conservation," *International Social Science Journal* 61, no. 199 (2010).

28. World People's Conference.

29. Immanuel Wallerstein, *World-Systems Analysis: An Introduction* (Durham, NC: Duke University Press, 2004).

30. Suzanne Romaine, "Planning for the Survival of Linguistic Diversity," *Language Policy* 5, no. 4 (2006); United Nations Secretariat of the Permanent Forum on Indigenous Issues, *State of the World's Indigenous Peoples* (New York: United Nations Publications, 2009).

31. Frank Waters, *The Book of Hopi: The First Revelation of the Hopi's Historical and Religious World-View of Life* (New York: Penguin Books, 1963; repr., 1971), 26.

32. Z. Brzezinski, *Strategic Vision: America and the Crisis of Global Power* (Philadelphia: Basic Books, 2012); see book description.

33. José Martínez Cobo, *Study of the Problem of Discrimination against Indigenous Populations* (New York: UN Economic and Social Council Commission on Human Rights, 1983).

34. Taiaiake Alfred and Jeff Corntassel, "Being Indigenous: Resurgences against Contemporary Colonialism," *Politics of Identity* 9 (2005).

35. Graeme Ward and Claire Smith, *Indigenous Cultures in an Interconnected World* (London: Allen and Unwin, 2000).

36. United Nations Secretariat of the Permanent Forum on Indigenous Issues.

37. See, for example, the work of James C. Scott, *The Art of Not Being Governed: An Anarchist History of Upland Southeast Asia* (New Haven: Yale University Press, 2009).

38. Richard Griggs, "The Meaning of 'Nation' and 'State' in the Fourth World," in *CWIS Occasional Paper* (Olympia, WA: Center for World Indigenous Studies, 1992).

39. Vine Deloria, *God Is Red: A Native View of Religion*, 30th anniversary ed. (Golden, CO: Fulcrum, 2003).

40. Quoted in Franke Wilmer, *The Indigenous Voice in World Politics: Since Time Immemorial* (Thousand Oaks, CA: Sage Publications, 1993), 117.

41. Peter Jacques, Sharon Ridgeway, and Richard Witmer, "Federal Indian Law and Environmental Policy: A Social Continuity of Violence," *Journal of Environmental Law and Litigation* 18, no. 2 (2003); Richard Robbins, "The Guarani: The Economics of Ethnocide," in *The Indigenous Experience: Global Perspectives*, eds. Roger Maaka and Chris Andersen, 150–157 (Toronto: Canadian Scholar's Press, 2006); Richard Robbins, *Global Problems and the Culture of Capitalism* (Upper Saddle River, NJ: Prentice Hall, 2010).

42. Francis Fukuyama, *The End of History and the Last Man* (New York: Free Press, 1992).

43. Robbie Robertson, *The Three Waves of Globalization: A History of a Developing Global Consciousness* (New York: Zed Books, 2003).

44. Ibid. See also Peter J. Jacques, *Globalization and the World Ocean* (Lanham, MD: AltaMira/Rowman and Littlefield, 2006).

45. Waters, 334.

46. See, for example, the explicit lesson in John Dearing, "Integration of World and Earth Systems: Heritage and Foresight," in *The World System and the Earth System*, eds. Alf Hornberg and Carole Crumley (Walnut Creek, CA: Left Coast Press, 2007).

47. Note, there are those who disagree and say that Easter Island was something of a utopia until British colonists introduced disease, slavery, and influenced maladaptive cultural practices, but both theories (ecological suicide or cultural imperialism) offer important lessons about civilization collapse. See Carl P. Lipo, T. L. Hunt, S. R. Haoa, "The 'Walking' Megalithic Statues (Moai) of Easter Island," *Journal of Archaeological Science*, 2012.

48. Just a few of the references that discuss various aspects of this claim are Ramachandra Guha and Juan Martinez-Alier, *Varieties of Environmentalism* (London: Earthscan, 1997);

H. G. Bohle, T. E. Downing, and M. J. Watts, "Climate Change and Social Vulnerability—Towards a Sociology and Geography of Food Insecurity," *Global Environmental Change* 4 (1994); W. Neil Adger, "Vulnerability," *Global Environmental Change* 15 (2006); R. J. Nicholls et al., "Coastal Systems and Low-Lying Areas," in *Climate Change 2007: Impacts, Adaptation and Vulnerability. Contribution of Working Group II to the Fourth Assessment Report of the Intergovernmental Panel on Climate Change,* eds. M. L. Parry et al. (Cambridge: Cambridge University Press, 2007); Parry et al., *Climate Change 2007: Impacts, Adaptation and Vulnerability. Contribution of Working Group II to the Fourth Assessment Report of the Intergovernmental Panel on Climate Change*; Giorgos Kallis, "Droughts," *Annual Review of Environment and Resources* 33, no. 1 (2008); B. L. Turner and Paul Robbins, "Land-Change Science and Political Ecology: Similarities, Differences, and Implications for Sustainability Science," *Annual Review of Environment and Resources* 33, no. 1 (2008); Samuel S. Myers and Jonathan A. Patz, "Emerging Threats to Human Health from Global Environmental Change," *Annual Review of Environment and Resources* 34, no. 1 (2009); Karen O'Brien et al., "Mapping Vulnerability to Multiple Stressors: Climate Change and Globalization in India," *Global Environmental Change* 14 (2004); Rodolfo Dirzo and Peter H. Raven, "Global State of Biodiversity and Loss," *Annual Review of Environment and Resources* 28, no. 1 (2003).

49. Thomas Hall and and James Fenelon, "The Futures of Indigenous Peoples: 9-11 and the Trajectory of Indigenous Survival and Resistance," *Journal of World Systems Research* 10, no. 1 (2004).

50. Amory Starr, *Naming the Enemy: Anti-Corporate Movements Confront Globalization* (London: Pluto Press, 2000).

51. Wallerstein.

52. William T. Hipwell, "A Deleuzian Critique of Resource-Use Management Politics in Industria," *The Canadian Geographer/Le Géographe Canadien* 48, no. 3 (2004).

53. Wallerstein, 24 (emphasis in original).

54. Ibid.

55. Ibid., 29.

56. Ibid., 65.

57. Ibid., 63.

58. Ibid., 75.

59. Hipwell.

60. Hipwell uses this in place of the more common "resource management" in order to keep the focus on the human use of natural resources; see 357.

61. Hipwell, 358.

62. Chris Mooney, "The Truth about Fracking," *Scientific American* 305, no. 5 (2011): 80–85.

63. William T. Hipwell, "The Industria Hypothesis: 'The Globalization of What?'" *Peace Review* 19, no. 3 (2007): 307.

64. Ibid.

65. Ibid.

66. Hipwell, 360, emphasis in original.

67. Ibid., 361.

68. Ibid., 372.

69. Paul J. Crutzen and J. Lelieveld, "Human Impacts on Atmospheric Chemistry," *Annual Review of Earth and Planetary Sciences* 29, no. 1 (2001): 17–45.

70. Peter Dauvergne, *The Shadows of Consumption: Consequences for the Global Environment* (Cambridge, MA: MIT Press, 2008).

Chapter Two

1. Oronto Douglas and Ike Okonta, "Ogoni People of Nigeria versus Big Oil," in *Paradigm Wars: Indigenous Peoples' Resistance to Globalization,* eds. Jerry Mander and Victoria Tauli-Corpuz (San Francisco: Sierra Club Books, 2006), 153.

2. Jon Frederick, *Empty Promises: The IMF, the World Bank, and the Planned Failures of Global Capitalism* (Washington, DC: IMF/World Bank 50 Years Is Enough Network, 2003), 5.

3. Catherine Caufield, *Masters of Illusion: The World Bank and the Poverty of Nations* (New York: Henry Holt, 1996), 34.

4. Ibid., 44.

5. Frederick, 4. See also William I. Robinson, "Global Capitalism Theory and the Emergence of Transnational Elites," *Critical Sociology* 38, no. 3 (2012); Robert Paehlke, *Democracy's Dilemma: Environment, Social Equity, and the Global Economy* (Cambridge, MA: MIT Press, 2004); Leslie Sklair, "Transnational Capitalist Class," in *The Wiley-Blackwell Encyclopedia of Globalization*, ed. George Ritzer (Hoboken, NJ: Blackwell Publishing, 2012). For evidence of this elite global capitalist class, see William I. Robinson and Jerry Harris, "Towards a Global Ruling Class? Globalization and the Transnational Capitalist Class," *Science and Society* 64, no. 1 (2000).

6. David Woodward, *The IMF and World Bank in the 21st Century: The Need for Change* (London: New Economic Foundation, 2005). Again, commentators argue we are in a "post–Bretton Woods period" but this refers to the collapse only of the policies that put capital and finance limits on currencies, not the demise of the actual institutions or their power to organize free trade as a global priority.

7. James Petras and Henry Veltmeyer, *Globalization Unmasked: Imperialism in the 21st Century* (New York: Zed Books, 2003), 5.

8. Quoted in Leslie Sklair, *Globalization: Capitalism and Its Alternatives*, 3rd ed. (New York: Oxford University Press, 2002), 49.

9. Caufield, 42.

10. Ibid., 197; see also Michael Breen, "IMF Conditionality and the Economic Exposure of Its Shareholders," *European Journal of International Relations* (September 6, 2012).

11. Woodward, 4.

12. Robert Wade, "U.S. Keeps Control of the World Bank," *Le Monde Diplomatique* (English ed.), October 24, 2012, http://mondediplo.com/blogs/us-keeps-controll-of-world-bank.

13. Caufield, 200.

14. Ibid., 197.

15. Ibid., 43.

16. Bank Information Center, "World Bank (IBRD & IDA)," www.bicusa.org/insitutions/worldbank#structure.

17. Caufield, 205.

18. Ibid., 65.

19. Ibid., 202.

20. Bretton Woods Project: Critical Voices on the World Bank and IMF, "IDA Replenishment," February 15, 2010, www.brettonwoodsproject.org/art-565921.

21. Ibid.

22. Rebecca Nelson, "Multilateral Development Banks: Overview and Issues for Congress," *Congressional Research Service*, March 7, 2011, 25.

23. Nick Beams, "German Nominee Forced to Withdraw," *World Socialist Web Site*, March 2000, https://wsws.org/articles/2000/mar2000/imf-m08.shtml.

24. Arvind Virmani, "Quota Formula Reform Is about IMF Credibility," *VOX*, June 22, 2012, www.voxeu.org/article/quota-formula-reform-about-imf-credibility.

25. Woodward, 4.

26. Michael DaCosta, "IMF Governance Reform and the Board's Effectiveness." *Global Policy* 3, no. 2 (2012): 242.

27. Bretton Woods Project, "Less Than Meets the Eye: IMF Reform Fails to Revolutionise the Institution," November 9, 2010, www.brettonwoodsproject.org/art-567128, 2.

28. DaCosta, 242.

29. Bernhard Steinki and Wolfgang Bergthaler, "Recent Reforms of the Finances of the

International Monetary Fund: An Overview," in *European Yearbook of International Economic Law*, eds. Christoph Herrmann and Jörg Philipp Terhechte (Berlin: Springer, 2012); Ayse Kaya, "Conflicted Principals, Uncertain Agency: The International Monetary Fund and the Great Recession," *Global Policy* 3, no. 1 (2012): 24–34.

30. Daniel Gros et al., "The Case for IMF Quota Reform," *Council on Foreign Relations*, October 11, 2012, www.cfr.org/imf/case-imf-quota-reform/, 2.

31. Nicholas G. Faraclas, "Melanesia, the Banks, and the Bingos: Real Alternatives Are Everywhere (Except in the Consultants' Briefcases)," in *There Is an Alternative: Subsistence and Worldwide Resistance to Corporate Globalization*, eds. Veronika Bennholdt-Thomsen, Nicholas Faraclas, and Claudia von Werlhof (New York: Zed Books, 2001), 71.

32. Sklair, *Globalization*, 47.

33. Michel Chossudovsky, *The Globalization of Poverty and the New World Order*, 2nd ed. (Pincourt, Quebec: Global Research, 2003), 5.

34. Quoted in Arturo Escobar, *Encountering Development: The Making and Unmaking of the Third World* (Princeton: Princeton University Press, 1995), 3.

35. Ibid., 5.
36. Ibid., 41.
37. Ibid., 216.
38. Ibid., 53.
39. Ibid., 8.

40. Ibid. Escobar draws on the work of Chandra Mohanty, who refers to the characterization of women in the Third World by feminists as having "needs" and "problems" but few choices and no freedom to act.

41. Ibid., 156.
42. Caufield, 43.
43. Escobar, 84.
44. Caufield, 56.
45. Ibid., 58.
46. Ibid.
47. Ibid., 59.
48. Ibid.
49. Ibid., 60.
50. Ibid., 61.
51. Ibid.
52. Escobar, 89.
53. Caufield, 242.
54. Ibid., 64.
55. Ibid., 71.
56. Ibid.
57. Ibid., 126.
58. Ibid., 137.

59. Amory Starr, *Naming the Enemy: Anti-Corporate Movements Confront Globalization* (London: Pluto Press, 2000), 15.

60. "An Examination of the Banking Crises of the 1980s and Early 1990s," in *An Examination of the Banking Crises* George Hanc (Washington, DC: Federal Deposit Insurance Corporation [FDIC], 1997), 191.

61. Walden Bello, "Building an Iron Cage," in *Views from the South: The Effects of Globalization and the WTO on Third World Countries*, ed. Sarah Anderson (Chicago: Food First Books and the International Forum on Globalization, 2000), 56.

62. Ibid., 57.
63. Ibid., 58.
64. Ibid., 65.
65. Starr, 16.

66. Escobar, 93.
67. Chossudovsky, xxi.
68. Ibid., 1.
69. Ibid., 20.
70. James Raymond Vreeland, *The IMF and Economic Development* (Cambridge: Cambridge University Press, 2003), 14.
71. Jerry Mander, "Foreword: The Resistance to Southern Perspectives," in Anderson, *Views from the South*, 2.
72. Ibid., 3.
73. Ibid.
74. Petras and Veltmeyer, 94.
75. Ibid., 97.
76. Ibid.
77. Ibid.
78. Ibid.
79. Elizabeth Stuart, *Blind Spot: The Continued Failure of the World Bank and IMF to Fully Assess the Impact of Their Advice on Poor People* (Oxford: Oxfam International, 2007).
80. Ibid., 2.
81. Structural Adjustment Participatory Review International Network (SAPRIN), *Structural Adjustment: The SAPRI Report, the Policy Roots of Economic Crisis, Poverty and Inequality* (New York: Zed Books, 2004), 4.
82. Ibid., 24.
83. Ibid., 25.
84. Ibid., 32.
85. Ibid., 23.
86. Ibid.
87. Ibid., 177.
88. Ibid., 190.
89. Ibid., 196.
90. Ibid., 199.
91. Ibid., 3.
92. Ibid., 213.
93. Ibid., 214.
94. Ibid., 40.
95. Ibid., 47.
96. Ibid., 55.
97. Sklair, *Globalization*, 127.
98. Ibid., 128–129.
99. "Structural Adjustment," 48.
100. Ibid., 50.
101. Ibid., 58.
102. Ibid., 48.
103. Stuart, 4.
104. Bretton Woods Project, "IMF Structural Conditionality Here to Stay," June 17, 2008, www.brettonwoodsproject.org/art-561809.
105. Axel Dreher and Martin Gassebner, "Do IMF and World Bank Programs Induce Government Crises? An Empirical Analysis," *International Organization* 66, no. 2 (2012): 329–358.
106. Breen, 1.
107. Andrea F. Presbitero and Alberto Zazzaro, "IMF Lending in Times of Crisis: Political Influences and Crisis Prevention," *World Development* 40, no. 10 (2012): 1944–1969.
108. Miguel A. Centeno and Joseph N. Cohen, "The Arc of Neoliberalism," *Annual Review of Sociology* 38, no. 1 (2012): 319.

162 NOTES

109. Ibid.; see also Clarence Y. H. Lo, "Countermovements and Conservative Movements in the Contemporary U.S.," *Annual Review of Sociology* 8 (1982): 107–134.
110. Centeno and Cohen, 320.
111. Ibid., 326.
112. Ibid.
113. Breen, 18.
114. Sklair, *Globalization,* 122.
115. Escobar, 57.
116. Chossudovsky, 5.
117. William T. Hipwell, "A Deleuzian Critique of Resource-Use Management Politics in Industria," *The Canadian Geographer/Le Géographe Canadien* 48, no. 3 (2004), 364.
118. Starr, 38.
119. Ibid., 12.

Chapter Three

1. Winona LaDuke, "Aspects of Traditional Knowledge and Worldview," in *Paradigm Wars: Indigenous Peoples' Resistance to Globalization,* eds. Jerry Mander and Victoria Tauli-Corpuz, eds. (San Francisco: Sierra Club Books, 2006), 24.
2. European Network on Debt and Development, "World Bank and IMF Emergency Loans: A Cure or Cause of the Food Crisis?" 2008, www.eurodad.org/whatsnew/articles.aspx?id=2402, June 18, 2008. Studies show a confidence rate of 99 percent that SAPs are an independent cause of infant mortality in sub-Saharan Africa. One reason is that, in order to satisfy the demands of SAPs, countries often devalue their currency, and this means it takes more money to buy the same food. For a family that spends 50 percent of their daily budget or more on food, this kind of change is devastating. See Carrie L. Shandra, John M. Shandra, and Bruce London, "The International Monetary Fund, Structural Adjustment, and Infant Mortality: A Cross-National Analysis of Sub-Saharan Africa," *Journal of Poverty* 16, no. 2 (2012). See also the discussion of Rosamond L. Naylor and Walter P. Falcon, "Food Security in an Era of Economic Volatility," *Population and Development Review* 36, no. 4 (2010).
3. John Vidal, "UN Warns of Looming Worldwide Food Crisis in 2013," *The Guardian,* October 13, 2012, www.guaradian.co.uk/global-development/2012/oct/14/un-warns.
4. Please see Chapter 1 for an in-depth discussion. See also Gordon Laxer and Dennis Soron, eds., *Not for Sale: Decommodifying Public Life* (Peterborough, Ontario: Broadview Press, 2006), 152.
5. Lori Wallach and Patrick Woodall, *Whose Trade Organization? A Comprehensive Guide to the WTO* (New York: The New Press, 2004), 4.
6. Nlerum S. Okogbule, "Globalization, Economic Sovereignty, and African Development: From Principles to Realities," *Journal of the Third World Studies* (Spring 2008): 3.
7. Ibid., 2.
8. For an in-depth view of dependency theory, see Fernando Henrique Cardoso and Enzo Faletto, *Dependency and Development in Latin America,* trans. Mariory Mattingly Urquidi (Berkeley: University of California Press, 1979). For more current understandings of the role of dependency in terms of trade arguments, see Immanuel Wallerstein, *World-Systems Analysis: An Introduction* (Durham, NC: Duke University Press, 2004), 12–14.
9. Andrew Mushita and Carol Thompson, *Biopiracy of Biodiversity: Global Exchange as Enclosure* (Trenton, NJ: Africa World Press 2007), 111.
10. Alliance for Democracy, "GATS: In Whose Service?," in *Empty Promises: The IMF, the World Bank and Planned Failures of Global Capitalism,* ed. IMF/World Bank 50 Years Is Enough Network (Washington, DC: IMF/World Bank 50 Years Is Enough Network, 2003), 52.

11. Ibid., 52.
12. Wallach and Woodall, 123.
13. Ibid., 124.
14. Alliance for Democracy, 52.
15. Ibid., 53.
16. Wallach and Woodall, 116. See also A. Claire Cutler, "Unthinking the GATS: A Radical Political Economy Critique of Private Transnational Governance," in *Business and Global Governance,* eds. Morten Ougaard and Anna Leander (New York/Abingdon, UK: Routledge, 2010).
17. Structural Adjustment Participatory Review International Network (SAPRIN), *Structural Adjustment: The SAPRI Report, the Policy Roots of Economic Crisis, Poverty and Inequality* (New York: Zed Books, 2004), 214.
18. Ibid., 214.
19. Ibid., 195; Shandra, Shandra, and London.
20. "Structural Adjustment," 196.
21. Ibid., 200. See also Shandra, Shandra, and London; K. Lee, "For Debate. The Impact of Globalization on Public Health: Implications for the UK Faculty of Public Health Medicine," *Journal of Public Health* 22, no. 3 (2000).
22. Wallach and Woodall, 125–126.
23. Jim Shultz, "Dignity and Defiance: Stories from Bolivia's Challenge to Globalization," eds. Jim Shultz and Melissa Crane Draper (Berkeley: University of California Press, 2008).
24. Todd Allee and Clint Peinhardt, "Delegating Differences: Bilateral Investment Treaties and Bargaining over Dispute Resolution Provisions," *International Studies Quarterly* 54, no. 1 (2010).
25. Cutler.
26. Wallach and Woodall, 63. Also see the discussion by Richard Peet, *Unholy Trinity: The IMF, World Bank, and WTO,* 2nd ed. (New York: Zed Books 2009), 211–212.
27. Wallach and Woodall, 63.
28. Ibid., 66.
29. Ibid., 69. See also B. D. Goldstein, "Toxicology and Risk Assessment in the Analysis and Management of Environmental Risk," in *Oxford Textbook of Public Health, Volume 2: The Methods of Public Health,* eds. R. Detels, R. Beaglehole, M. A. Lansing, and M. Gulliford (Oxford: Oxford University Press, 2009); Michael Balter, "Scientific Cross-Claims Fly in Continuing Beef War," *Science* 284, no. 5419 (1999).
30. Wallach and Woodall, 20.
31. Ibid., 57. See also G. Skogstad, "Internationalization, Democracy, and Food Safety Measures: The (Il)Legitimacy of Consumer Preferences," *Global Governance* 7 (2001).
32. Wallach and Woodall, 20.
33. Ibid., 30–31.
34. Peet, 212. See also Christophe Bonneuil and Les Levidow, "How Does the World Trade Organization Know? The Mobilization and Staging of Scientific Expertise in the GMO Trade Dispute," *Social Studies of Science* 42, no. 1 (2012).
35. Investment measures have been forcefully resisted by developing nations within the WTO but are being increasingly added to regional free trade agreements such as NAFTA.
36. Wallach and Woodall, 21.
37. Ibid., 42.
38. Ibid.
39. Robert Hunter Wade, "What Strategies Are Viable for Developing Countries Today? The World Trade Organization and the Shrinking of 'Development Space,'" *Review of International Political Economy* 10, no. 4 (2003), 621.
40. Vandana Shiva, *Earth Democracy: Justice, Sustainability, and Peace* (Cambridge, MA: South End Press, 2005), 1.
41. Quoted in ibid., 1.
42. Shiva, 21.

164 NOTES

43. Ibid., 40. See also Millennium Ecosystem Assessment (MEA), *Ecosystems and Human Well-Being: Our Human Planet: Summary for Decision-Makers* (Washington, DC: Island Press, 2005).

44. See, for example, Ramachandra Guha, *The Unquiet Woods: Ecological Change and Peasant Resistance in the Himalayas* (Berkeley: University of California Press, 2000); Ramachandra Guha and Juan Martinez-Alier, *Varieties of Environmentalism* (London: Earthscan, 1997).

45. Mushita and Thompson, 61.
46. Quoted in ibid., 75.
47. Ibid., 63.
48. Mushita and Thompson, 21.
49. Ibid., 22.
50. Shiva, 146.
51. Wallach and Woodall, 205.
52. Mushita and Thompson, 68.
53. Wade.
54. Mushita and Thompson, 64.
55. Ibid., 10.
56. Ibid., 4.

57. Martin Khor, "WTO and the Third World: On a Catastrophic Course," *Third World Traveler*, October/November 1999, www.thirdworldtraveler.com/WTO_MAI/, 5.

58. Ibid., 6.
59. Ibid.
60. Ibid., 8.

61. Wallach and Woodall, 193. See also Timothy A. Wise, "Promise or Pitfall? The Limited Gains from Agricultural Trade Liberalisation for Developing Countries," *Journal of Peasant Studies* 36, no. 4 (2009).

62. Mushita and Thompson, 24.
63. Ibid., 118.
64. Ibid., 119.
65. Ibid., 120.

66. See, for example, David Blandford et al., "How Effective Are WTO Disciplines on Domestic Support and Market Access for Agriculture?" *The World Economy* 33, no. 11 (2010).

67. Wallach and Woodall, 195–196. See also Blandford et al.

68. Naty Bernadino, "Ten Years of the WTO Agreement of Agriculture: Problems and Prospects," paper presented at the WTO Public Symposium, Switzerland, April 20–22, 2005; "WTO after 10 Years: Global Problems and Multilateral Solutions," www.wto.org/engilsh/news_e/events_smpo5_e/bernadino15_e/pdf (panel sponsored by the International Gender and Trade Network–Asia, 2005), 2.

69. Kimberly Ann Elliott, "Delivering on Doha: Farm Trade and the Poor" (Washington, DC: Center for Global Development, 2006), 18.

70. Wallach and Woodall, 196. See also H. G. Jensen and H. Zobbe, "Consequences of Reducing Limits on Aggregate Measurement of Support," in *Agricultural Trade Reform and the Doha Development Agenda*, eds. K. Anderson and W. Martin (Washington, DC/Basingstoke, UK: World Bank/Palgrave Macmillan, 2006).

71. Chakravarthi Raghavan, "Seattle WTO Ministerial Ends in failure," *Third World Network*, 1999, www.twnside.org.sg/title/deb2-cn.htm.

72. Wallach and Woodall, 7.
73. Elliott, 1.
74. Ibid., 2.

75. Martin Khor, "Trade: Rules Meeting Postponed, New Text Won't Be Issued," *Third World Network*, 2008, www.twnside.org.sg/title2/wto.info20080811.htm.

76. Lori Wallach, "RIP Doha Round" *Public Citizen*, 2008.

77. Martin Khor, "Agriculture: Falconer Report Says WTO Talks Failed on Political Factor," *Third World Network,* 2008, www.ytnside.org.sg/title2/resurgence/215/cover4.doc, 2.
78. Ibid., 3.
79. Deborah James, *Globalization: Leaving the WTO Behind,* Center for Economic and Policy Research, 2008, 1.
80. Tom Miles, "WTO Sees Tentative Progress in Trade Talks after Stalemate," *Reuters,* October 3, 2012, www.reuters.com/assets/print?aid =USL 6E8L3HWW20122003.
81. Walden Bello, "How to Manufacture a Global Food Crisis: Lessons from the World Bank, IMF and WTO," *Focus on the Global South,* 2008, http://focusweb.org/search/node/slsden%soBello%20.
82. Ibid.
83. United Nations Food and Agriculture Organization.
84. David. S. Battisti and Rosamond L. Naylor, "Historical Warnings of Future Food Insecurity with Unprecedented Seasonal Heat," *Science* 323, no. 5911 (January 9, 2009): 240–244.
85. Wolfram Schlenker and Michael J. Roberts, "Nonlinear Temperature Effects Indicate Severe Damages to U.S. Crop Yields under Climate Change," *PNAS* 106, no. 37 (September 15, 2009): 15594–15598; see also Michael J. Roberts, Wolfram Schlenker, and Jonathan Eyer, "Agronomic Weather Measures in Econometric Models of Crop Yield with Implications for Climate Change," *American Journal of Agricultural Economics* (May 22, 2012): aas047.

Chapter Four

1. Daniel R. Wildcat and Vine Deloria Jr., *Power and Place: Indian Education in America* (Golden, CO: Fulcrum Resources, 2001), 70.
2. Immanuel Wallerstein, *World-Systems Analysis: An Introduction* (Durham, NC: Duke University Press, 2004), 22. We recognize the critiques that Wallerstein pays too little attention to culture, and acknowledge that culture is resistant to change. However, we also have presented our argument for the existence of human agency and the ability to acknowledge cultural constraints.
3. Nicholas G. Faraclas, "Melanesia, the Banks, and the BINGOs: Real Alternatives Are Everywhere (Except in the Consultants' Briefcases)," in *There Is an Alternative: Subsistence and Worldwide Resistance to Corporate Globalization,* eds. Veronika Bennholdt-Thomsen, Nicholas Faraclas, and Claudia von Werlhof (New York: Zed Books, 2001), 67.
4. Walden Bello, "Capitalism's Crisis and Our Response," in *Conference on the Global Crisis Sponsored by Die Linke Party and Rosa Luxemburg Foundation* (Berlin: Transnational Institute, 2009), www.tri.org/archives/bello_crisisandresponse.
5. Ibid., 2.
6. Ibid., 3.
7. Ibid.
8. Wallerstein, 2.
9. Ibid., 49.
10. Robert Oak, "House Passed Bill Which Closes the 'Offshore Outsourcing' International Corporate Tax Scheme," *Economic Populist,* 2010, 2.
11. Ibid., 80.
12. Chye-Ching Huang and Chand Stone, *Putting U.S. Corporate Taxes in Perspective,* Center on Budget and Policy Priorities, 2008, 1.
13. Ibid.
14. Stephen G. Bunker and Paul S. Ciccantell, *Globalization and the Race for Resources* (Baltimore: John Hopkins University Press, 2005), xii.
15. Robin Broad and John Cavanagh, *Development Redefined: How the Market Met Its Match* (Boulder: Paradigm Publishers, 2009), 3.

16. Ibid., 3.
17. Wallerstein, 86.
18. Ibid., 86. Chase-Dunn, Kwon, Lawrence, and Inoue also show that a turn to more intensive financialization and credit lines to bring capital into the core are some signs of a declining hegemon, as in the decline of the nineteenth-century British Empire.
19. Christopher Chase-Dunn, Roy Kwon, Kirk Lawrence, and Hiroko Inoue, "Last of the Hegemons: U.S. Decline and Global Governance," *International Review of Modern Sociology* 37, no. 1 (2011): 1–29; see also Alf Hornberg and Carole Crumley, eds., *The World System and the Earth System: Global Socioenvironmental Change and Sustainability since the Neolithic* (Walnut Creek, CA: Left Coast Press, 2007); Jackie Smith and Dawn Wiest, *Social Movements in the World-System: The Politics of Crisis and Transformation* (New York: Russell Sage Foundation, 2012); Christopher Chase-Dunn and Thomas Hall, *Rise and Demise: Comparing World Systems* (Boulder: Westview Press, 1997); Christopher Chase-Dunn, "Social Evolution and the Future of World Society," *Journal of World Systems Research* 11, no. 2 (2005): 171–194.
20. Peter Custers, *Capital Accumulation and Women's Labor in Asian Economies* (London: Zed Books, 1997).
21. Rosa Luxemburg, quoted in June C. Nash, *Mayan Visions: The Quest for Autonomy in an Age of Globalization* (New York: Routledge, 2001), 6.
22. Ibid., 6.
23. Silvia Federici, "War, Globalization and Reproduction," in Bennholdt-Thomsen, Faraclas, and von Werlhof, *There Is an Alternative*, 133.
24. Perhaps the definitive discussion on hegemony is Christopher Chase-Dunn et al., "Hegemony and Social Change," *Mershon International Studies Review* 38, no. 2 (1994): 361–376.
25. Peter J. Jacques and Jessica Racine Jacques, "Monocropping Cultures into Ruin: The Loss of Food Varieties and Cultural Diversity," *Sustainability* 4, no. 10 (2012): 2970–2997.
26. Gerry Marten, "'Non-Pesticide Management' for Agricultural Pests: Escaping the Pesticide Trap—Controlling Pests without Chemical Pesticides Catapults Farmers from Chronic Poisoning and Debt to Health and Hope," *The Eco-Tipping Points Project*, 2012, www.ecotippingpoints.org/our-stories/indepth/india-pest-management-nonpesticide-neem.html.
27. Maria Mies and Veronika Bennholdt-Thomsen, *The Subsistence Perspective: Beyond the Globalised Economy* (New York: Zed Books, 1999), 3.
28. Maria Mies, "Towards a Methodology for Feminist Research," 1983, quoted in ibid., 20.
29. Mies and Bennholdt-Thomsen, 34, emphasis in original.
30. Angela Hilmi, *Agricultural Transition: A Different Logic* (Oslo: The More and Better Network, 2012); María Elena Martínez-Torres and Peter M. Rosset, "La Vía Campesina: The Birth and Evolution of a Transnational Social Movement," *Journal of Peasant Studies* 37, no. 1 (January 1, 2010): 149–175. See also Jacques and Jacques, appendix 1.
31. Janet M. Conway, *Edges of Global Justice: The World Social Forum and Its "Others"* (New York: Routledge, 2013), 10.
32. Gilles Deleuze and Felix Guattari, *A Thousand Plateaus: Capitalism and Schizophrenia* (Minneapolis: University of Minnesota Press, 1987).
33. William Hipwell, "A Deleuzian Critique of Resource-Use Management Politics in Industria," *The Canadian Geographer/Le Géographe Canadien* 48, no. 3 (2004): 356–377.
34. Conway.
35. Janet Conway, "Cosmopolitan or Colonial? The World Social Forum as 'Contact Zone,'" *Third World Quarterly* 32, no. 2 (2011): 224.
36. Ibid., 225.
37. Boaventura De Sousa Santos, "The World Social Forum and the Global Left," *Politics and Society* 36, no. 2 (June 1, 2008): 251.
38. J. C. Scott, *The Moral Economy of the Peasant: Rebellion and Subsistence in Southeast Asia* (New Haven: Yale University Press, 1977).

39. Eric Hobsbawm, *Age of Empire 1875–1914* (London: Weidenfeld and Nicolson, 2010).
40. Adam David Morton, "Global Capitalism and the Peasantry in Mexico: The Recomposition of Class Struggle," *Journal of Peasant Studies* 34, no. 3–4 (July 1, 2007): 441–473.
41. Mies and Bennholdt-Thomsen, 86.
42. Ibid.
43. Ibid., 87.
44. Stephen Gudeman, "Vital Energy: The Current of Relations," *Social Analysis* 56, no. 1 (2012): 57–73.
45. Stephan Gudeman, *The Anthropology of Economy: Community, Market, and Culture* (Malden, MA: Blackwell Publishers, 2001), 7.
46. Ibid., 37.
47. Ibid., 99–100.
48. Ibid., 102.
49. Ibid., 110.
50. Ibid., 118.
51. Ibid.
52. Nash, 26.
53. Ibid., 27.
54. Ibid., 25.
55. Ibid., 122.
56. Ibid., 145.
57. Ana Cecilia Dinerstein and Séverine Deneulin, "Hope Movements: Naming Mobilization in a Post-Development World," *Development and Change* 43, no. 2 (2012): 585–602.
58. Ibid., 590.
59. Nash, 147.
60. Ibid.
61. "The International Peasant's Voice," *La Via Campesina*, 2011, http://viacampesina.org/en/.
62. Peter Rosset, "Food Sovereignty and the Contemporary Food Crisis," *Development* 51, no. 4 (2008); Peter Rosset, "The Time Has Come for La Via Campesina and Food Sovereignty," *Focus on the Global South*, 2008, http//focusweb.org/node/1874.
63. Amartya Kumar Sen, *Poverty and Famines: An Essay on Entitlement and Deprivation* (New York: Oxford University Press, 1981).
64. Raj Patel, *Stuffed and Starved: Markets, Power and the Hidden Battle for the World Food System* (London: Portebello Books, 2007), 130.
65. Rosset.
66. Ibid., 2.
67. *World Bank and IMF Emergency Loans: A Cure or Cause of the Food Crisis?* (Brussels, Belgium: European Network on Debt and Development, 2008). See also P. A. Sánchez et al., *Halving Hunger: It Can Be Done* (London/Sterling VA: United Nations Millennium Project Task Force on Hunger, 2005). The Millennium Project estimates the numbers break down along the following lines of malnourished people: 50 percent are small holder farmers who cannot grow or buy enough for their household; 20 percent are landless poor; and 10 percent are pastoralists, fishers, and those who subsist in/on forests.
68. Russell Mokhiber and Robert Weissman, "Corporate Globalization and the Poor," *New Renaissance Magazine*, www.ru.org/econotes. The original report is in Mark Weisbrot, Robert Naiman, and Joyce Kim, *The Emperor Has No Growth: Declining Economic Growth Rates in the Era of Globalization* (Center for Economic and Policy Research, 2000). Corroborated by Branko Milanovic, "The Two Faces of Globalization: Against Globalization as We Know It," *World Development* 31, no. 4 (2003).
69. Mokhiber and Weissman.
70. Rosset.

71. Walden Bello, "How to Manufacture a Global Food Crisis: Lessons from the World Bank, IMF and WTO," *Focus on the Global South*, 2008, http://focusweb.org/search/node/Wlsden%20Bello%20.

72. See, for example, Douglas Southgate, Douglas Graham, and Luther Tweeten, *The World Food Economy* (Malden, MA: Wiley-Blackwell, 2007); R. Paarlberg, *Food Politics: What Everyone Needs to Know* (New York: Oxford University Press, 2010).

73. "Europe and Gobal Food Security," EurActiv, 2009, www.euractiv.com/en/cap/europe-global-food-security-linksdossier-188488.

74. Rachel Kerr, "Lessons from the Old Green Revolution for the New: Social, Environmental and Nutritional Issues for Agricultural Change in Africa," *Progress in Development Studies* 12 (2012): 219; see also the meticulous research of Joseph Cotter, *Troubled Harvest: Agronomy and Revolution in Mexico, 1880–2002* (Westport, CT: Praeger, 2003).

75. Vandana Shiva, *The Violence of the Green Revolution: Third World Agriculture, Ecology and Politics*, 4th ed. (New York: Zed Books, 2000), 31.

76. Ibid., 128.

77. Ibid., 197.

78. See, for example, Prabhu Pingali, "Green Revolution: Impacts, Limits, and the Path Ahead," *Proceedings of the National Academy of Sciences* 109 (2012): 12302–12308. (Pingali is a representative of the Bill and Melinda Gates Foundation.)

79. Ibid., 200; T. Kastner, M. J. I. Rivas, W. Koch, and S. Nonhebel, "Global Changes in Diets and the Consequences for Land Requirements for Food," *Proceedings of the National Academy of Sciences* 109 (2012): 6868–6872.

80. Pingali, 12304; Kerr.

81. Ibid., 200.

82. Vandana Shiva, "Globalization and Poverty," in Bennholdt-Thomsen, Faraclas, and von Werlhof, *There Is an Alternative*, 57.

83. Swarup Dutta, "Green Revolution Revisited: The Contemporary Agrarian Situation in Punjab, India," *Social Change* 42 (2012): 231; Food First, "The Gates-Rockefeller Green Revolution for Africa: Still Ignoring the Root Causes of Hunger?" September 18, 2006, www.foodfirst.org/node1506; B. B. Mohanty, "'We Are Like the Living Dead': Farmer Suicides in Maharashtra, Western India," *Journal of Peasant Studies* 32, no. 2 (2005).

84. Patel, 25.

85. Ibid., 26.

86. A. Sarkar, "Sustaining Livelihoods in Face of Groundwater Depletion: A Case Study of Punjab, India," *Environment, Development, and Sustainability* 14 (2012): 183–195; B. Manjunatha, D. Hajong, and R. Koulagi, "Impact of Agricultural Development on Environment," *Environment and Ecology* 30 (2012): 393–396; P. Srivastava, M. Gupta, and S. Mukherjee, "Mapping Spatial Distribution of Pollutants in Groundwater of a Tropical Area of India Using Remote Sensing and GIS," *Applied Geomatics* 4 (2012): 21–32; G. Kaur, R. S. Sawhney, and R. Vohra, "Micro-Cantilever Sensors for Detection of Pesticide Contents in the Water-Table of Malwa Region in Punjab," *International Journal of Advanced Research in Computer Engineering and Technology* 1 (2012): 285–290.

87. Food First.

88. Ibid.

89. Chander Suta Dogra, "Poison Earth," *Outlook India*, December 24, 2007, www.outlookindia.com/article.aspx?236312.

90. Ibid.

91. Ibid.

92. Shiva, *The Violence of the Green Revolution*, 185.

93. Shiva, "Globalization and Poverty," 58. See also Jules Pretty, "Agricultural Sustainability: Concepts, Principles and Evidence," *Philosophical Transactions of the Royal Society B: Biological Sciences* 363, no. 1491 (2008): 447–465. In the widest study ever conducted on agroecological approaches, Pretty shows that incorporating ecological principles, like

enhancing biological and genetic diversity or encouraging beneficial biological interactions (e.g., birds of prey for catching rodents), used by peasant agriculture (with some modern integration) increased yields 79 percent on average, and 25 percent of the areas in the study saw a doubling of yields.

94. Ibid., 58.
95. George Pyle, *Raising Less Corn, More Hell: The Case for the Independent Farm and against Industrial Food* (New York: PublicAffairs/Perseus Books Group, 2005), 65.
96. Andrew Mushita and Carol Thompson, *Biopiracy of Biodiversity: Global Exchange as Enclosure* (Trenton, NJ: Africa World Press, 2007), 85.
97. Patel, 2.
98. Ibid., 12.
99. Ibid., 2.
100. Ibid., 106.
101. Ibid., 38.
102. Ibid., 2.
103. Pyle, 84.
104. Ibid., 72.
105. Ibid., 73.
106. Ibid., 51.
107. Ibid., 92.
108. Ibid., 49.
109. Mushita and Thompson, 4.
110. Ibid., 7.
111. Ibid., 160.
112. Ibid., 158.
113. Ibid., 160.
114. Carol B. Thompson, "Alliance for a Green Revolution in Africa (AGRA): Advancing the Theft of African Genetic Wealth," *Review of African Political Economy* 39 (2012): 346, emphasis added. See also Eric Holt-Giménez, "Out of AGRA: The Green Revolution Returns to Africa," *Development* 51 (2008): 464–471.
115. *Voices from Africa: African Farmers and Environmentalists Speak Out against a New Green Revolution in Africa* (Oakland, CA: The Oakland Institute, 2009), http://media.oaklandinstitute.org/sites/oaklandinstitute.org/files/voicesfromafrica_full.pdf, 6.
116. Bello, 6.
117. Leahy.
118. Thompson, 346.
119. *Voices from Africa*, 1, emphasis added.
120. Leahy.
121. A speech at a Swissaid meeting in Berne, May 29, 2008. Henry Saragih, "Food Sovereignty to Answer World Food and Energy Crisis," *La Via Campesina*, June 18, 2008, http://viacampesina.org/en/index-php/mainmenu-27/biodiversity-and-genetic-resources.
122. Mariam Mayet, "The New Green Revolution in Africa: A Trojan Horse for GMOs," in *Voices from Africa: African Farmers and Environmentalists Speak Out against a New Green Revolution in Africa*, eds. Anuradha Mittal and Melissa Moore (Oakland, CA: The Oakland Institute, 2009), 14.
123. Amy Harmon and Andrew Pollack, "Battle Brewing over Genetically Modified Food," *New York Times*, May 24, 2012.
124. Mushita and Thompson, 41.
125. Ibid., 42.
126. Mayet, 2.
127. Ibid., 3.
128. Rosset.
129. Rodale Institute, *The Farming Systems Trial: Celebrating 30 Years*, 2012,

http://66.147.224.123; see also E. Holt-Giménez, A. Shattuck, M. Altieri, H. Herren, and S. Gliessman, "We Already Grow Enough Food for 10 Billion People ... and Still Can't End Hunger," *Journal of Sustainable Agriculture* 36 (2012): 595–598.

130. Tim LaSalle, Paul Hepperly, and Amadou Diop, *The Organic Green Revolution*, (Kutztown, PA: The Rodale Institute, 2008). While controversial, academic research supports this claim: see Catherine Badgley et al., "Organic Agriculture and the Global Food Supply," *Renewable Agriculture and Food Systems* 22, no. 2 (2007); R. Hine et al., *Organic Agriculture and Food Security in Africa* (New York: United Nations, 2008). See also Pretty; David Pimentel et al., "Environmental, Energetic, and Economic Comparisons of Organic and Conventional Farming Systems," *Bioscience* 55, no. 7 (2005).

131. Stephan Raspe, in Mies and Bennholdt-Thomsen, *The Subsistence Perspective*, 81.

132. See, for example, the demands of the More and Better Network: Angela Hilmi, *Agricultural Transition: A Different Logic* (Oslo: The More and Better Network, 2012).

133. M. Marcos Terena, "Native Peoples and the World Social Forum," Inter Press Service News Agency, 2008, www.forumsocialmundial.org.br/noticias_textos.php?cd_news=443.

Chapter Five

Portions of this chapter are reproduced from Peter J. Jacques, "Ecology, Distribution, and Identity in the World Politics of Environmental Skepticism," *Capitalism, Nature, Socialism* 19: 8–27, with permission from Taylor and Francis Publishers.

1. David Held, ed., "Central Perspectives on the Modern State," in *Political Theory and the Modern State* (Cambridge: Polity Press, 1989).

2. Donald Worster, *Rivers of Empire: Water, Aridity, and the Growth of the American West* (Newbury Park, CA: Sage/Pantheon Books, 1985), 262.

3. Walt W. Rostow, *Stages of Economic Growth: A Non-Communist Manifesto* (Cambridge: Cambridge University Press, 1960); A. F. K. Organski, *The Stages of Political Development* (New York: Knopf, 1965).

4. Joseph Cotter, *Troubled Harvest: Agronomy and Revolution in Mexico, 1880–2002*, Contributions in Latin American Studies (Westport, CT: Praeger, 2003); Swarup Dutta, "Green Revolution Revisited: The Contemporary Agrarian Situation in Punjab, India," *Social Change* 42, no. 2 (June 1, 2012): 229–247; Rachel Bezner Kerr, "Lessons from the Old Green Revolution for the New: Social, Environmental and Nutritional Issues for Agricultural Change in Africa," *Progress in Development Studies* 12, no. 2/3 (2012): 213–229.

5. Miguel A. Centeno and Joseph N. Cohen, "The Arc of Neoliberalism," *Annual Review of Sociology* 38, no. 1 (2012): 317–340. Of course, the Soviet Union and its socialist bloc would oppose neoliberalism until it dissolved in 1991, when the former Soviet Union became one of the most strident neoliberal countries under the leadership of Boris Yeltsin.

6. Friedrich von Hayek, *The Road to Serfdom* (Chicago: University of Chicago Press, 1944); Milton Friedman, *Capitalism and Freedom* (Chicago: University of Chicago Press, 1962).

7. For a summary of anarchist thinking, see Peter Marshall, *Demanding the Impossible: A History of Anarchism* (Oakland, CA: PM Press, 2010); Jamie Peck and Adam Tickell, "Apparitions of Neoliberalism: Revisiting 'Jungle Law Breaks Out,'" *Area* 44, no. 2 (2012): 245–249; on regular bailouts of big capital during regular crises, see Centeno and Cohen.

8. Jamie Peck and Adam Tickell, "Neoliberalizing Space," *Antipode* 34, no. 3 (2002): 385.

9. Rajeev C. Patel, "Food Sovereignty: Power, Gender, and the Right to Food," *Public Library of Science Medicine* 9, no. 6 (2012): e1001223; see also Carol Thompson, "Alliance for a Green Revolution in Africa (AGRA): Advancing the Theft of African Genetic Wealth," *Review of African Political Economy* 39, no. 132 (2012): 345–350.

10. T. Alfred, "Sovereignty," in *A Companion to American Indian History*, eds. Philip J. Deloria and Neal Salisbury (Oxford: Blackwell Publishing, 2007): 460–474.

11. Karl Polanyi, *The Great Transformation* (New York: Beacon Press, 1957).
12. Ibid., 57.
13. William Hipwell, "A Deleuzian Critique of Resource-Use Management Politics in Industria," *The Canadian Geographer/Le Géographe Canadien* 48, no. 3 (2004): 367.
14. Ibid., 361–362.
15. Ibid., 368.
16. Ibid., 370.
17. James C. Scott, "State Simplifications: Nature, Space and People," *Journal of Political Philosophy* 3, no. 3 (1995): 191–233. James C. Scott, *Seeing Like a State: How Certain Schemes to Improve the Human Condition Have Failed* (Yale University Press, 1998).
18. James C. Scott, *Two Cheers for Anarchism: Six Easy Pieces on Autonomy, Dignity, and Meaningful Work and Play* (Princeton: Princeton University Press, 2012).
19. Franke Wilmer, *The Indigenous Voice in World Politics: Since Time Immemorial* (Thousand Oaks, CA: Sage Publications, 1993); Thomas Hall and James Fenelon, "The Futures of Indigenous Peoples: 9-11 and the Trajectory of Indigenous Survival and Resistance," *Journal of World Systems Research* 10, no. 1 (2004).
20. David Maybury-Lewis, *Indigenous Peoples, Ethnic Groups, and the State* (Boston: Allyn and Bacon, 1996). 4.
21. Frank Waters, *The Book of Hopi: The First Revelation of the Hopi's Historical and Religious World-View of Life* (New York: Penguin Books, 1963; repr., 1971).
22. Chase D. Mendenhall, Gretchen C. Daily, and Paul R. Ehrlich, "Improving Estimates of Biodiversity Loss," *Biological Conservation* 151, no. 1 (2012): 32–34; Rodolfo Dirzo and Peter H. Raven, "Global State of Biodiversity and Loss," *Annual Review of Environment and Resources* 28, no. 1 (2003); Peter M. Vitousek et al., "Human Domination of Earth's Ecosystems," *Science* 277 (1997).
23. Gabriela Kütting, *Globalization and the Environment: Greening Global Political Economy* (Albany: State University of New York Press, 2004); Thomas Princen, Michael Maniates, and Ken Conca, *Confronting Consumption* (Cambridge, MA: MIT Press, 2002).
24. Wilmer, 5.
25. Immanuel Wallerstein, *The Modern World-System I: Capitalist Agriculture and the Origins of the European World-Economy in the Sixteenth Century* (Berkeley: University of California Press, 2011), 136.
26. Ibid., 133.
27. Patricia K. Townsend, *Environmental Anthropology: From Pigs to Policies*, 2nd ed. (Groveland, IL: Waveland Press, 2009); Patrick V. Kirch, "Archaeology and Global Change: The Holocene Record," *Annual Review of Environment and Resources* 30, no. 1 (2005).
28. See, for example, Robyn Eckersley, *The Green State: Rethinking Democracy and Sovereignty* (Cambridge: MA: MIT Press. 2004).
29. Obviously, Marxist theories are meant to be revolutionary, not mainstream or dominant, but we refer to them in this way here to refer to the triad of explanations that dominate ideas about the state.
30. Held.
31. Antonio Gramsci, *Prison Notebooks*, Volumes 1–3, ed. J. A. Buttigieg (New York: Columbia University Press, 1996).
32. For a good review of theories of the state, see Murray Knutilla and Wendee Kubik, *State Theories: Classical, Global and Feminist Perspectives* (Halifax, Nova Scotia: Fernwood Publishing, 2000).
33. Emile Durkheim, *The Division of Labor in Society* (New York: The Free Press, 1950 [1893]).
34. Weber, quoted inKnutilla and Kubik, 50, emphasis in original.
35. Georg Wilhelm Friedrich Hegel, *Hegel's Philosophy of Right*, trans. T. M. Knox (Oxford: Clarendon Press, 1965 [1821]).
36. Peter Jacques, Sharon Ridgeway, and Richard Witmer, "Federal Indian Law and

Environmental Policy: A Social Continuity of Violence," *Journal of Environmental Law and Litigation* 18, no. 2 (2003).

37. The 1648 Peace of Westphalia at the end of the 30 Years War in Europe is sometimes considered the clearest beginning of the modern nation-state because the treaty included agreements for noninterference in other countries' internal affairs and other elements familiar to the modern state system. However, the history presented in this chapter puts the emergence of the nation-state at least as early as the sixteenth century.

38. Melissa Burchard, "Returning to the Body" (PhD Dissertation, University of Minnesota, 1996).

39. Jean-Jacques Rousseau, *The Social Contract* (New York: Penguin, 1968 [1762]).

40. Thomas Hobbes, "Introduction," *Leviathan*, ed. C. B. Macpherson (New York: Penguin, 1985 [1651]).

41. Ibid.

42. Francis Paul Prucha, *The Great Father: The United States Government and the American Indians* (Lincoln: University of Nebraska Press, 1984).

43. Held, 25–26.

44. See, for example, Marx's illustration of this in Karl Marx and Friedrich Engels, eds., *The Eighteenth Brumaire of Louis Bonaparte* (Salt Lake City, UT: Project Gutenberg Press, 2006 [1852]).

45. Ralph Miliband, *The State in Capitalist Society* (New York: Basic Books, 1969); Nicos Poulantzas, "The Problem of the Capitalist State," in *Ideology and Social Science*, ed. R. Blackburn (London: Fontana, 1972).

46. Held, 28.

47. Ibid.

48. Ibid.

49. Durkheim.

50. Poulantzas.

51. Antonio Gramsci, *Selections from the Prison Notebooks of Antonio Gramsci*, eds. Quintin Hoare and Geoffrey Nowell Smith (London: Lawrence and Wishart, 1971).

52. Anthony Giddens, *The Constitution of Society: The Outline of the Theory of Structuration* (Cambridge: Polity Press, 1984).

53. Eckersley.

54. John Barry and Robyn Eckersley, eds., *The State and the Global Ecological Crisis* (Cambridge, MA: MIT Press, 2005); John Dryzek et al., *Green States and Social Movements: Environmentalism in the United States, United Kingdom, Germany, and Norway* (Oxford: Oxford University Press, 2003).

55. David John Frank, Ann Hironaka, and Evan Schofer, "The Nation-State and the Natural Environment over the Twentieth Century," *American Sociological Review* 65, no. 1 (2000).

56. Eckersley.

57. Murray Bookchin, "Society and Ecology," in *Debating the Earth: The Environmental Politics Reader*, eds. John Dyzek and David Schlosberg (New York: Oxford University Press, 1998), 424.

58. John Barry, *Rethinking Green Politics* (Thousand Oaks, CA: Sage Publications, 1999), 93.

59. Matthew Paterson, *Understanding Global Environmental Politics: Domination, Accumulation and Resistance* (New York: St. Martin's Press, 2000), 66.

60. Ibid., 70.

61. Worster.

62. David Held et al., *Global Transformations: Politics, Economics and Culture* (Stanford: Stanford University Press, 1999).

63. Ibid., 39.

64. Ibid.

65. Richard Bean, "War and the Birth of the Nation State," *The Journal of Economic History* 33, no. 1 (1973): 203–221.

66. Ibid., 41, emphasis added.
67. Ibid.
68. Peter J. Jacques, *Globalization and the World Ocean* (Lanham: MD: AltaMira/Rowman and Littlefield, 2006).
69. Ibid.
70. Walter Rodney, *How Europe Underdeveloped Africa* (Washington, DC: Howard University Press, 1982).
71. Wallerstein.
72. Hannah Arendt, *The Human Condition* (Chicago: University of Chicago Press, 1958).
73. Stephen Bunker and and Paul Ciccantell, eds., "Economic Ascent and the Global Environment: World-Systems Theory and the New Historical Materialism," in *Ecology and the World-System*, eds. W. Goldfrank, D. Goodman, and A. Szasz (Westport, CT: Greenwood Press, 1999), 107.
74. Stephen Bunker, *Underdeveloping the Amazon: Extraction, Unequal Exchange, and the Failure of the Modern State* (Chicago: University of Chicago Press, 1985), 21.
75. Ibid., 21–22.
76. Anthony Giddens, *Central Problems in Social Theory: Action, Structure, and Contradiction in Social Analysis* (Berkeley: University of California Press, 1979).
77. David Schlosberg, "Reconceiving Environmental Justice: Global Movements and Political Theories," *Environmental Politics* 13, no. 3 (2004).
78. For an application of Marxist and Weberian theories of the state to the Third World, where he finds that these theories are indeed useful, see Adrien Leftwich, "States of Underdevelopment: The Third World State in Perspective," *Journal of Theoretical Politics* 6, no. 1 (1994).

Chapter Six

1. Quoted in Shannon Biggs, "From the Mouths of Babes at Rio+20 'Join Me in Earth Revolution,'" *Global Exchange,* June 28, 2012, www.globalexchange.org/blogs/peopletopeople/2012/06/28/from-the-mouths-of-babes-at-rio20-join-me-in-earth-revolution/.
2. Immanuel Wallerstein, *World-Systems Analysis: An Introduction* (Durham, NC: Duke University Press, 2004).
3. Jackie Smith and Dawn Wiest, *Social Movements in the World-System: The Politics of Crisis and Transformation*, American Sociological Association's Rose Series in Sociology (New York: Russell Sage Foundation, 2012), 6.
4. Ibid., 5.
5. Ibid.
6. Roberto De Vogli, *Progress or Collapse: The Crises of Market Greed* (New York: Routledge, 2013), book description.
7. John L. Casti, *X-Events: The Collapse of Everything* (New York: HarperCollins, 2012), 4.
8. Ibid.
9. Indigenous Peoples Global Conference On Rio+20 and Mother Earth, "Kari-Oca 2 Declaration," Rio de Janeiro, Brazil, 2012, www.ienearth.org/docs/DECLARATION-of-KARI-OCA-2-Eng.pdf.
10. Makere Stewart-Harawira, "Returning the Sacred: Indigenous Ontologies in Perilous Times," in *Radical Human Ecology: Intercultural and Indigenous Approaches,* eds. Lewis Williams, Rose Roberts, and Alastair McIntosh (Farnham UK: Ashgate, 2012), 73–88.
11. See the fascinating work of Suzanna J. Opree, Moniek Buijzen, and Patti M. Valkenburg, "Lower Life Satisfaction Related to Materialism in Children Frequently Exposed to Advertising," *Pediatrics* 130, no. 3 (September 1, 2012): e486–e491. See also Daniel Kahneman, Allen B. Krueger, David Schkade, Norbert Schwarz, and Arthur A. Stone, "Would You Be Happier If You Were Richer? A Focusing Illusion," *Science* 312, no. 5782 (2006): 1908–1910.

12. The classic works on the modernization theory of development are Abramo Fimo Kenneth Organski, *The Stages of Political Development* (New York: Knopf, 1965); and Walt W. Rostow, *Stages of Economic Growth: A Non-Communist Manifesto* (Cambridge: Cambridge University Press, 1960).

13. World Conference of Indigenous Peoples on Territory, Environment and Development, "Indigenous Peoples Earth Charter and Kari-Oca Declaration," Kari Oca, Brazil, 1992, www.trc.org.nz/sites/trc.org.nz/files/Indigenous%20Peoples%20Earth%20Charter.pdf, emphasis added.

14. For example, in *The Fountainhead*, a brilliant architect is hindered by corrupt corporatism (symbolic of socialism). He is able to shine when let loose from the binds of personal relations, and the main love interest castigates herself for loving beauty (in art) and the architect himself. But, who raised this brilliant architect? Who educated this architect, and how did he come to the privilege of this education? Rand would have us think that we are truly free only inasmuch as we do everything for ourselves, making the capitalist selfishness a perverted virtue. Neither truth nor joy exists in such stark Manichean terms of the Self versus Other or individualism versus community. Ayn Rand, *The Fountainhead* (New York: Bobbs-Merrill Company, 1971).

15. Douglas Keay, "Interview for Woman's Own," in *Thatcher Archive* (London: Margaret Thatcher Foundation, 1987), emphasis added. The illogical condition of the phrase, "If children have a problem, it is society that is at fault. There is no such thing as society," where a problem belongs to something, society, that supposedly does not exist, is illustrative of the requirement to solve social problems with social action; therefore, by Thatcher's own rhetoric and definition, society must exist.

16. Paul-Marie Boulanger, "The Life-Chances Concept: A Sociological Perspective in Equity and Sustainable Development," in *Sustainable Development: Capabilities, Needs, and Well-Being*, eds. Felix Rauschmayer, Ines Omann, and Johannes Fruhmann (New York: Routledge, 2011), 83–103.

17. Friedrich Engels, "The Great Towns," in *The City Reader*, eds. Richard T. LeGates and Frederic Stout, Urban Reader Series (New York: Routledge, 1996).

18. Robert Chernomas and Ian Hudson, "Social Murder: The Long-Term Effects of Conservative Economic Policy," *International Journal of Health Services* 39, no. 1 (2009): 107.

19. Gary Weiss, *Ayn Rand Nation: The Hidden Struggle for America's Soul* (New York: St. Martin's Press, 2012), 4.

20. Although the financial industry is highly regulated in the sense that the states demand they observe webs of bureaucratic conditions, such as disclosures to mortgage applicants, the financial sectors across the world-system are unregulated in the sense that they have enormous discretion to do what they want, often with open impunity to the public good.

21. Andrew Tangel and Alejandro Lazo, "JP Morgan Targeted by U.S. Task Force for Role in Financial Mess," *Los Angeles Times*, October 2, 2012, online.

22. Weiss, 3–4.

23. Andrew Ross Sorkin, "JP Morgan's $12 Billion Bailout," *The New York Times Blog*, March 18, 2008.

24. Edmund L. Andrews, "Greenspan Concedes Error on Regulation," *New York Times*, October 23, 2008.

25. "'Shadow Banking' up to $67 Trillion, Financial Group Says," *New York Times*, November 19, 2012, B6.

26. Roger Keil, "Sustaining Modernity, Modernizing Nature: The Environmental Crisis and the Survival of Capitalism," in *The Sustainable Development Paradox: Urban Political Economy in the United States and Europe*, eds. Rob Krueger and David Gibbs (New York: Guilford Press, 2007), 45.

27. See the interesting discussion of William Ophuls, *Plato's Revenge: Politics in the Age of Ecology* (Cambridge, MA: MIT Press, 2011).

28. Aldo Leopold, *A Sand County Almanac with Essays on Conservation from Round River*, ed. R. Finch (New York: Oxford University Press/Ballantine Books Edition, 1970 [1949]), 197.

29. Rodolfo Dirzo and Peter H. Raven, "Global State of Biodiversity and Loss," *Annual Review of Environment and Resources* 28, no. 1 (2003); Peter M. Vitousek et al., "Human Domination of Earth's Ecosystems," *Science* 277 (1997).

30. See the warnings of Johan Rockström et al., "A Safe Operating Space for Humanity," *Nature* 461, no. 7263 (2009); Johan Rockström et al., "Planetary Boundaries: Exploring the Safe Operating Space for Humanity," *Ecology and Society* 14, no. 2 (2009); Will Steffen, Johan Rockström, and Robert Costanza, "How Defining Planetary Boundaries Can Transform Our Approach to Growth," *Solutions* 2, no. 3 (2011); Graham M. Turner, "A Comparison of the Limits to Growth with 30 Years of Reality," *Global Environmental Change* 18, no. 3 (2008); Donnella Meadows, Jorgen Randers, and Dennis Meadows, *Limits to Growth: The 30-Year Update* (White River Junction, VT: Chelsea Green Publishing, 2004); Mathis Wackernagel et al., "Tracking the Ecological Overshoot of the Human Economy," *Proceedings of the National Academy of Science* (*PNAS*) 99, no. 14 (2002).

31. Makere Stewart-Harawira, *The New Imperial Order: Indigenous Responses to Globalization* (New York: Zed Books, 2005), 32, emphasis added.

32. Sharon Venne, "She Must Be Civilized: She Paints Her Toe Nails," in *A Will to Survive: Indigenous Essays on the Politics of Culture, Language, and Identity,* ed. Stephen Greymorning (New York: McGraw-Hill, 2004), 130.

33. Val Plumwood, *Feminism and the Mastery of Nature* (New York: Routledge, 1993).

34. Victor Montejo, "Mayan Ways of Knowing: Modern Mayans and the Elders," in *The Will to Survive: Indigenous Essays on the Politics of Culture, Language, and Identity,* ed. Stephen Greymorning (New York: McGraw-Hill, 2004), 168.

35. Vine Deloria, *Custer Died for Your Sins: An Indian Manifesto* (New York: Macmillan, 1969).

36. Ward Churchill, "A Question of Identity," in *A Will to Survive: Indigenous Essays on the Politics of Culture, Language, and Identity,* ed. Stephen Greymorning (New York: McGraw-Hill, 2004).

37. Sharon O'Brien, *American Indian Tribal Governments,* The Civilization of the American Indian (Norman: University of Oklahoma Press, 1989), 14.

38. Ibid., 14–15.

39. It may be duly noted that tribes around the world are impoverished and in many examples corrupt. For example, in the United States, some modern tribal governments have been accused of violence and other crimes against their people. Though circumstances vary, many of these pathologies are attributable to imposed changes from settler governments meant to "civilize" tribes and make them resemble Western governing norms or other impositions from mainstream culture that create pathologies when they push out traditions. These demands have changed traditional governing practices, which allow members of the tribe to exploit the new system for their own gain, often in conjunction for outside gain, such as coal companies who lease land with modern tribal governments who gained power against the traditional processes. In the United States, American Indian scholars indicate that tribal corruption can often be traced to Western liberal standards that were imposed on tribes in the 1930s that forced tribal governments to use majority-rule elections. Tribal governance was almost always consensual not majoritarian, and majoritarian elections went against the traditional decision-making processes that had been in place for eons. For example, in traditional Hopi culture, silence was a form of dissent. If one is silent in an election, his/her dissent is ignored, and opportunistic individuals in the tribes are then able to win elections handily. See, for example, David E. Wilkins and Heidi K. Stark, *American Indian Politics and the American Political System,* 3rd ed. (Lanham, MD: Rowman and Littlefield, 2010).

40. David Rich Lewis, "Native Americans and the Environment: A Survey of Twentieth-Century Issues," *American Indian Quarterly* 19, no. 3 (1995): 423, emphasis added.

41. This section focuses more on Western Hemisphere Indigenous societies because these are the voices most accessible in print resources from the United States.

42. Mason Durie, "Indigenous World Views and Therapeutic Pathways," in *Encounter in*

Pastoral Care and Spiritual Healing: Towards an Integrative and Intercultural Approach, eds. D. J. Louw, Takaaki David Ito, and Ulrike Elsdörfer (Zurich: LIT Verlag, 2012), 10.

43. Stewart-Harawira.

44. Quoted in Makere Stewart-Harawira, "Returning the Sacred: Indigenous Ontologies in Perilous Times," in *Radical Human Ecology: Intercultural and Indigenous Approaches,* 82.

45. Stewart-Harawira, *The New Imperial Order,* 39.

46. Ibid.

47. Durie in ibid., 39–40.

48. Stewart-Harawira, "Returning the Sacred," 84, emphasis added.

49. Montejo, 165.

50. Ibid., 164.

51. Ibid., 165.

52. Joseph A. Tainter, *The Collapse of Complex Societies* (Cambridge: Cambridge University Press, 1988).

53. O'Brien, 27.

54. Dennis Wall and Virgil Masayesva, "People of the Corn: Teachings in Hopi Traditional Agriculture, Spirituality, and Sustainability," *American Indian Quarterly* 28, no. 3/4 (2004): 435–436.

55. Ibid., 435–436.

56. O'Brien, 30.

57. Ibid., 31.

58. Sitting Bull in ibid., 147.

59. Ibid.

60. Akim D. Reinhardt, *Ruling Pine Ridge: Oglala Lakota Politics from the Ira to Wounded Knee* (Lubbock: Texas Tech University Press, 2007), 205. GOON was an acronym for Guardians of the Oglala Nation and was openly used by both sides.

61. See Clark's preface to Leonard Peltier, *Prison Writings: My Life Is My Sun Dance* (New York: St. Martin's Griffin, 1999).

62. O'Brien.

63. Chris Hedges and Joe Sacco, *Days of Destruction, Days of Revolt* (New York: Nation Books/Perseus Academic, 2012), 8. The authors note that rape of Lakota women was a regular practice by US soldiers, but now "The fury of self-destruction sweeps across Pine Ridge like the Black Plague.... The violence imposed on Indian culture has been internalized. Despair and pain of this magnitude lead to lives dedicated to self-immolation" (4).

64. M. Scott Taylor, "Buffalo Hunt International Trade and the Virtual Extinction of the North American Bison," *American Economic Review* 101, no. 7 (2006): 3162–3195.

65. Osage/Cherokee scholar George Tinker cautions that the word "God" is usually inadequate to explain Wakan and other Native American ideas of the "Sacred Other," where "God" is a gloss over the complexities of sacred elements. Indian tribal spiritualities would refer to a "spiritual force that permeates the whole world and is manifest in countless ways in the world around us at any given moment and especially in any given place" (157). Tinker also notes that the Judeo-Christian prejudice is to think that God has a special preference for humanity over the rest of existence, and that this "runs the risk of generating human arrogance, which too easily sees the world in terms of hierarchies of existence, all of which are ultimately subservient to the needs and whims of humans" (156). See George Tinker, "An American Indian Theological Response to Eco-Justice," in *Defending Mother Earth: Native American Perspectives on Environmental Justice,* ed. Jace Weaver (Maryknoll, NY: Orbis Books, 1996).

66. Jeffrey Ostler, "'They Regard Their Passing as Wakan': Interpreting Western Sioux Explanations for the Bison's Decline," *The Western Historical Quarterly* 30, no. 4 (1999): 479–480.

67. Ibid., 480, emphasis added.

68. Excerpt from Arvol Looking Horse, *White Buffalo Teachings from Chief Arvol Looking Horse, 19th Generation Keeper of the Sacred White Buffalo Pipe of the Lakota, Dakota and

Nakota Great Sioux Nation (Williamsburg, MA: Dreamkeepers Press, 2001), found at his website: http://arvollookinghorse.homestead.com/Chief_Arvol_Excerpts_Book_1.html.
69. Tinker, 158.
70. LaDuke, quoted in Deborah McGregor, "Coming Full Circle: Indigenous Knowledge, Environment, and Our Future," *American Indian Quarterly* 28, no. 3/4 (2004): 393–394.
71. Ibid.

Chapter Seven

1. Arvol Looking Horse, *White Buffalo Teachings from Chief Arvol Looking Horse, 19th Generation Keeper of the Sacred White Buffalo Pipe of the Lakota, Dakota and Nakota Great Sioux Nation* (Williamsburg, MA: Dreamkeepers Press, 2001), quoted from his website: http://arvollookinghorse.homestead.com/Chief_Arvol_Excerpts_Book_1.html.
2. David Rich Lewis, "Native Americans and the Environment: A Survey of Twentieth-Century Issues," *American Indian Quarterly* 19, no. 3 (1995).
3. Robert J. Diaz and Rutger Rosenberg, "Spreading Dead Zones and Consequences for Marine Ecosystems," *Science* 321, no. 5891 (2008).
4. Ibid., 926.
5. Joseph Cotter, *Troubled Harvest: Agronomy and Revolution in Mexico, 1880–2002*, Contributions in Latin American Studies (Westport, CT: Praeger, 2003), 4.
6. Peter M. Vitousek et al., "Human Domination of Earth's Ecosystems," *Science* 277 (1997).
7. Bradley Reed Howard, *Indigenous Peoples and the State: The Struggle for Native Rights* (Dekalb: Northern Illinois Press, 2003), 3.
8. See also Thomas Hall and James Fenelon, "The Futures of Indigenous Peoples: 9-11 and the Trajectory of Indigenous Survival and Resistance," *Journal of World Systems Research* 10, no. 1 (2004).
9. International Indian Treaty Council/Consejo Internacional de Tratados Indios, "Declaration of Continuing Independence by the First International Indian Treaty Council" (International Indian Treaty Council, 1974), emphasis in original.
10. "Kari-Oca Declaration," *Kari-Oca Conference* (Auckland: Conference of Churches in Aotearoa New Zealand, 1992), signed on May 30, 1992, reaffirmed on June 4, 2002, in Bali, Indonesia.
11. "Indigenous People's Earth Charter," in *Kari-Oca Conference*.
12. International Indigenous Peoples Summit on Sustainable Development, "The Kimberley Declaration of the International Indigenous Peoples Summit on Sustainable Development," 2002, www.iwgia.org/sw217.asp, emphasis added.
13. Ibid., emphasis added.
14. See Lorie M. Graham and Siegfried Wiessner, "Indigenous Sovereignty, Culture, and International Human Rights Law," *South Atlantic Quarterly* 110, no. 2 (2011).
15. International Indigenous Peoples Summit on Sustainable Development
16. Indigenous Peoples Global Conference on Rio+20 and Mother Earth, "Kari-Oca 2 Declaration," Rio de Janeiro, Brazil, 2012, http://indigenous4motherearthrioplus20.org/kari-oca-2-declaration/.
17. "Indigenous People's Earth Charter."
18. George Tinker, "An American Indian Theological Response to Eco-Justice," in *Defending Mother Earth: Native American Perspectives on Environmental Justice*, ed. Jace Weaver (Maryknoll, NY: Orbis Books, 1996), 160.
19. Ibid.
20. Ibid., 162.
21. Ibid., 163.
22. Peter Dauvergne, *The Shadows of Consumption: Consequences for the Global Environment*

(Cambridge, MA: MIT Press, 2008); Thomas Princen, *The Logic of Sufficiency* (Cambridge, MA: MIT Press, 2005).

23. Ramachandra Guha, *The Unquiet Woods: Ecological Change and Peasant Resistance in the Himalaya* (Berkeley: University of California Press, 2000).

24. Andrew Szasz, *Shopping Our Way to Safety: How We Changed from Protecting the Environment to Protecting Ourselves* (Minneapolis: University of Minnesota Press, 2007).

25. See the devastating work of Mike Davis, *Planet of Slums* (New York: Verso Books, 2007).

26. World Social Forum, *Action Objectives* (World Social Forum, 2008), emphasis in original.

27. Paul Polak, *Out of Poverty* (San Francisco: Berret-Koehler, 2008).

28. Ivan Illich, *Energy and Equity* (San Francisco: Calder and Boyars, 1974).

29. Ibid., 29.

30. See S. Greymorning, "Culture and Language: The Political Realities to Keep Trickster at Bay," *Canadian Journal of Native Studies* 20, no. 1 (2000); Angela Cavender Wilson, "Reclaiming Our Humanity: Decolonization and the Recovery of Indigenous Knowledge," in *War and Border Crossings: Ethics When Cultures Clash,* eds. Peter A. French and Jason A. Short (Oxford: Rowman and Littlefield, 2005); Luisa Maffi, "Endangered Languages, Endangered Knowledge," *International Social Science Journal* 54, no. 173 (2002); Luisa Maffi, "Linguistic, Cultural, and Biological Diversity," *Annual Review of Anthropology* 34 (2005); Luisa Maffi, "Bio-Cultural Diversity for Endogenous Development: Lessons from Research, Policy, and On-the-Ground Experiences," paper presented at the conference Endogenous Development and Bio-Cultural Diversity: The Interplay of Worldviews Globalization and Locality, Geneva, Switzerland, October 3–5, 2006.

31. Jessica Gordon, "Manifesto," *The Official Idle No More Site,* January 2013, http://idlenomore.ca/manifesto; "Idle No More: Is This Just the Beginning? First Nation History-Makers Weigh In on Idle's Future, Where Non-Aboriginals Fit In, and What Would Tecumseh Do Anyway?" *Now,* Toronto, February 2012, www.nowtoronto.com/news/story.cfm?content=191127.

32. Kurt Bayer, "Indigenous Protest Movement Spread to NZ," *New Zealand Herald,* Auckland, January 2013, www.nzherald.co.nz/nz/news/article.cfm?c_id=1&objectid=10857061.

33. Found in K. D. Moore and M. P. Nelson, *Moral Ground: Ethical Action for a Planet in Peril* (San Antonio, TX: Trinity University Press, 2010); Moore and Nelson's work led to the Blue River Quorum, "The Blue River Declaration: An Ethic for the Earth," in *Spring Creek Project for Ideas, Nature, and the Written Word* (Oregon State University, 2011).

34. "The Solutions," listed in Indigenous Peoples Global Conference on Rio+20 and Mother Earth.

35. Excerpts from Arvol Looking Horse.

REFERENCES

Adger, W. Neil. "Vulnerability." *Global Environmental Change* 15 (2006): 268–281.
Alfred, Taiaiake, and Jeff Corntassel. "Being Indigenous: Resurgences against Contemporary Colonialism." *Politics of Identity* 9 (2005): 597–614.
Allee, Todd, and Clint Peinhardt. "Delegating Differences: Bilateral Investment Treaties and Bargaining over Dispute Resolution Provisions." *International Studies Quarterly* 54, no. 1 (2010): 1–26.
Alliance for Democracy. "GATS: In Whose Service?" In *Empty Promises: The IMF, the World Bank and Planned Failures of Global Capitalism,* ed. IMF/World Bank 50 Years Is Enough Network. Washington, DC: IMF/World Bank 50 Years Is Enough Network, 2003.
Arendt, Hannah. *The Human Condition.* Chicago: University of Chicago Press, 1958.
Badgley, Catherine, et al. "Organic Agriculture and the Global Food Supply." *Renewable Agriculture and Food Systems* 22, no. 2 (2007): 86–108.
Bailey, Ronald. "The Slave(ry) Trade and the Development of Capitalism in the United States: The Textile Industry in New England." *Social Science History* 14, no. 3 (1990): 373–414.
Barry, John. *Rethinking Green Politics.* Thousand Oaks, CA: Sage Publications, 1999.
Barry, John, and Robyn Eckersely, eds. *The State and the Global Ecological Crisis.* Cambridge, MA: MIT Press, 2005.
Battisti, David. S., and Rosamond L. Naylor. "Historical Warnings of Future Food Insecurity with Unprecedented Seasonal Heat." *Science* 323, no. 5911 (January 9, 2009): 240–244.
Beams, Nick. "German Nominee Forced to Withdraw." *World Socialist Website.* https://wsws.org/articles/2000/mar2000/imf-m08.shtml.
Bean, Richard. "War and the Birth of the Nation State." *The Journal of Economic History* 33, no. 1 (1973): 203–221.
Bello, Walden. "Building an Iron Cage." In *Views from the South: The Effects of Globalization and the WTO on Third World Countries,* ed. Sarah Anderson. 54–94. Chicago, IL: Food First Books and the International Forum on Globalization, 2000.
———. "Capitalism's Crisis and Our Response." In *Conference on the Global Crisis*

Sponsored by Die Linke Party and Rosa Luxemburg Foundation. Berlin: Transnational Institute, 2009. www.tri.org/archives/bello_crisisandresponse.

———. "How to Manufacture a Global Food Crisis: Lessons from the World Bank, IMF and WTO." *Focus on the Global South,* 2008. http://focusweb.org/search/node/slsden%soBello%20.

Bernadino, Naty. "Ten Years of the WTO Agreement of Agriculture: Problems and Prospects." International Gender and Trade Network-Asia. Paper presented at the WTO Public Symposium in Switzerland, "WTO after 10 Years: Global Problems and Multilateral Solutions," April 20–22, 2005. www.wto.org/english/news_smpo5_e/bernadino15_e/pdf.

Biggs, Shannon. "From the Mouths of Babes at Rio+20: 'Join Me in Earth Revolution.'" *Global Exchange.* www.globalexchange.org/blogs/peopletopeople/2012/06/28/from-the-mouths-of-babes-at-rio20-join-me-in-earth-revolution/.

Blandford, David, Ivar Gaasland, Roberto Garcia, and Erling Vårdal. "How Effective Are WTO Disciplines on Domestic Support and Market Access for Agriculture?" *The World Economy* 33, no. 11 (2010): 1470–1485.

Blue River Quorum. "The Blue River Declaration: An Ethic for the Earth." *Spring Creek Project for Ideas, Nature, and the Written Word.* 2011. http://springcreek.oregonstate.edu/documents/BlueRiverDeclaraton.2012.pdf.

Bohle, H. G., T. E. Downing, and M. J. Watts. "Climate Change and Social Vulnerability—Towards a Sociology and Geography of Food Insecurity." *Global Environmental Change* 4 (1994): 37–48.

Bonneuil, Christophe, and Les Levidow. "How Does the World Trade Organization Know? The Mobilization and Staging of Scientific Expertise in the GMO Trade Dispute." *Social Studies of Science* 42, no. 1 (February 1, 2012): 75–100.

Bookchin, Murray. "Society and Ecology." In *Debating the Earth: The Environmental Politics Reader,* eds. John Dyzek and David Schlosberg. 415–428. New York: Oxford University Press, 1998.

Boulanger, Paul-Marie. "The Life-Chances Concept: A Sociological Perspective in Equity and Sustainable Development." In *Sustainable Development: Capabilities, Needs, and Well-Being,* eds. Felix Rauschmayer, Ines Omann, and Johannes Fruhmann. 83–103. Abingdon, UK: Routledge, 2011.

Bretton Woods Project. "IMF Structural Conditionality Here to Stay." www.brettonwoodsproject.org/art-561809.

———. "Less Than Meets the Eye: IMF Reform Fails to Revolutionise the Institution." November 9, 2010. www.brettonwoodsproject.org/art-567128.

Broad, Robin, and John Cavanagh. *Development Redefined: How the Market Met Its Match.* Boulder: Paradigm Publishers, 2009.

Brosius, J. P., and S. L. Hitchner. "Cultural Diversity and Conservation." *International Social Science Journal* 61, no. 199 (2010): 141–168.

Brzezinski, Zbigniew. *Strategic Vision: America and the Crisis of Global Power.* Philadelphia: Basic Books, 2012.

Bunker, Stephen. *Underdeveloping the Amazon: Extraction, Unequal Exchange, and the Failure of the Modern State.* Chicago: University of Chicago Press, 1985.

Bunker, Stephen, and Paul Ciccantell. "Economic Ascent and the Global Environment: World-Systems Theory and the New Historical Materialism." In *Ecology and the World-System,* eds. W. Goldfrank, D. Goodman, and A. Szasz. Westport, CT: Greenwood Press, 1999.

———. *Globalization and the Race for Resources*. Baltimore: The John Hopkins University Press, 2005.
Burchard, Melissa. "Returning to the Body." PhD dissertation, University of Minnesota, 1996.
Cardoso, Fernando Henrique, and Enzo Faletto. *Dependency and Development in Latin America*. Translated by Mariory Mattingly Urquidi. Berkeley: University of California Press, 1979.
Casti, John L. *X-Events: The Collapse of Everything*. New York: HarperCollins, 2012.
Caufield, Catherine. *Masters of Illusion: The World Bank and the Poverty of Nations*. New York: Henry Holt, 1996.
Centeno, Miguel A., and Joseph N. Cohen. "The Arc of Neoliberalism." *Annual Review of Sociology* 38, no. 1 (2012): 317–340.
Chapin, F. Stuart, III, et al. "Consequences of Changing Biodiversity." *Nature* 405, no. 6783 (2000): 234–242.
Chase-Dunn, Christopher. "Social Evolution and the Future of World Society." *Journal of World Systems Research* 11, no. 2 (2005): 171–194.
Chase-Dunn, Christopher, and Thomas Hall. *Rise and Demise: Comparing World Systems*. Boulder: Westview Press, 1997.
Chase-Dunn, Christopher, et al. "Hegemony and Social Change." *Mershon International Studies Review* 38, no. 2 (1994): 361–376.
Chase-Dunn, Chris, Roy Kwon, Kirk Lawrence, and Hiroko Inoue. "Last of the Hegemons: U.S. Decline and Global Governance." *International Review of Modern Sociology* 37, no. 1 (2011): 1–29.
Chernomas, Robert, and Ian Hudson. "Social Murder: The Long-Term Effects of Conservative Economic Policy." *International Journal of Health Services* 39, no. 1 (2009): 107–121.
Chossudovsky, Michel. *The Globalization of Poverty and the New World Order*. 2nd ed. Pincourt, Canada: Global Research, 2003.
Churchill, Ward. "A Question of Identity." In *A Will to Survive: Indigenous Essays on the Politics of Culture, Language, and Identity*, ed. Stephen Greymorning. 59–94. New York: McGraw-Hill, 2004.
Clémençon, Raymond. "Welcome to the Anthropocene: Rio+20 and the Meaning of Sustainable Development." *The Journal of Environment and Development* 21, no. 3 (September 1, 2012): 311–338.
Cobo, José Martínez. *Study of the Problem of Discrimination against Indigenous Populations*. New York: UN Economic and Social Council Commission on Human Rights, 1983. UN Document: UN Commission on Human Rights, March 11, 1986. www.refworld.org/docid/3b00f02630.html.
Conway, Janet. "Cosmopolitan or Colonial? The World Social Forum as 'Contact Zone.'" *Third World Quarterly* 32, no. 2 (2011): 217–236.
———. *Edges of Global Justice: The World Social Forum and Its "Others."* New York: Routledge, 2013.
Cotter, Joseph. *Troubled Harvest: Agronomy and Revolution in Mexico, 1880–2002*. Contributions in Latin American Studies. Westport, CT: Praeger, 2003.
Crutzen, P. J., and J. Lelieveld. "Human Impacts on Atmospheric Chemistry." *Annual Review of Earth and Planetary Sciences* 29, no. 1 (2001): 17–45.
Custers, Peter. *Capital Accumulation and Women's Labor in Asian Economies*. London: Zed Books, 1997.

Cutler, A. Claire. "Unthinking the GATS: A Radical Political Economy Critique of Private Transnational Governance." In *Business and Global Governance*, eds. Morten Ougaard and Anna Leander. 78–96. New York: Routledge, 2010.
DaCosta, Michael. "IMF Governance Reform and the Board's Effectiveness." *Global Policy* 3, no. 2 (2012): 242–244.
Dauvergne, Peter. *The Shadows of Consumption: Consequences for the Global Environment*. Cambridge, MA: MIT Press, 2008.
Davis, Mike. *Planet of Slums*. New York: Verso Books, 2007.
Dearing, John. "Integration of World and Earth Systems: Heritage and Foresight." In *The World System and the Earth System*, eds. Alf Hornberg and Carole Crumley. 38–55. Walnut Creek, CA: Left Coast Press, 2007.
Deloria, Vine. *Custer Died for Your Sins: An Indian Manifesto*. London: Macmillan, 1969.
———. "Traditional Technology." In *Power and Place: Indian Education in America*, eds. Vine Deloria and Daniel R. Wildcat. 57–65. Golden, CO: Fulcrum Resources, 2001.
———. *God Is Red: A Native View of Religion*. 30th anniversary ed. Golden, CO: Fulcrum, 2003.
Diaz, Robert J., and Rutger Rosenberg. "Spreading Dead Zones and Consequences for Marine Ecosystems." *Science* 321, no. 5891 (August 15, 2008): 926–929.
Dirzo, Rodolfo, and Peter H. Raven. "Global State of Biodiversity and Loss." *Annual Review of Environment and Resources* 28, no. 1 (2003): 137–167.
Dobson, Andrew. "Listening: The New Democratic Deficit." *Political Studies* 60 (2012): 843–859.
Dogra, Chander Suta. "Poison Earth." *Outlook India*, December 24, 2007.
Donaldson, Laura E. "Writing the Talking Stick: Alphabetic Literacy as Colonial Technology and Postcolonial Appropriation." *American Indian Quarterly* 22, no. 1/2 (Winter/Spring 1998).
Douglas, Oronto, and Ike Okonta. "Ogoni People of Nigeria versus Big Oil." In *Paradigm Wars: Indigenous Peoples' Resistance to Globalization*, eds. Jerry Mander and Victoria Tauli-Corpuz. 153–155. San Francisco: Sierra Club Books, 2006.
Dreher, Axel, and Martin Gassebner. "Do IMF and World Bank Programs Induce Government Crises? An Empirical Analysis." *International Organization* 66, no. 2 (2012): 329–358.
Dryzek, John, et al. *Green States and Social Movements: Environmentalism in the United States, United Kingdom, Germany, and Norway*. Oxford: Oxford University Press, 2003.
Durie, Mason. "Indigenous World Views and Therapeutic Pathways." In *Encounter in Pastoral Care and Spiritual Healing: Towards an Integrative and Intercultural Approach*, eds. D. J. Louw, Takaaki David Ito, and Ulrike Elsdörfer. 9–17. Zurich: LIT Verlag, 2012.
Durkheim, Emile. *The Division of Labor in Society*. New York: The Free Press, 1950 [1893].
Dutta, S. "Green Revolution Revisited: The Contemporary Agrarian Situation in Punjab, India." *Social Change* 42 (2012): 229–247.
Eckersley, Robyn. *The Green State: Rethinking Democracy and Sovereignty*. Cambridge, MA: MIT Press, 2004.
Elliott, Kimberly Ann. *Delivering on Doha: Farm Trade and the Poor*. Washington, DC: Center for Global Development, 2006.
Engels, Friedrich. "The Great Towns." In *The City Reader*, eds. Richard T. LeGates and Frederic Stout. Urban Reader Series. 46–55. New York: Routledge, 1996.

Escobar, Arturo. *Encountering Development: The Making and Unmaking of the Third World*. Princeton: Princeton University Press, 1995.
"Europe and Global Food Security." EurActiv. www.euractiv.com/en/cap/europe-global-food-security-linksdossier-188488.
"An Examination of the Banking Crises of the 1980s and Early 1990s." In *An Examination of the Banking Crises*. Washington, DC: Federal Deposit Insurance Corporation (FDIC), 1997.
Faraclas, Nicholas G. "Melanesia, the Banks, and the BINGOS: Real Alternatives Are Everywhere (Except in the Consultants' Briefcases)." In *There Is an Alternative: Subsistence and Worldwide Resistance to Corporate Globalization*, eds. Veronika Bennholdt-Thomsen, Nicholas Faraclas, and Claudia von Werlhof. 67–76. New York: Zed Books, 2001.
Federici, Silvia. "War, Globalization and Reproduction." In *There Is an Alternative: Subsistence and Worldwide Resistance to Corporate Globalization*, eds. Veronika Bennholdt-Thomsen, Nicholas Faraclas, and Claudia von Werlhof. 133–145. New York: Zed Books, 2001.
Food First. "The Gates-Rockefeller Green Revolution for Africa: Still Ignoring the Root Causes of Hunger?" August 23, 2007. www.foodfirst.org/node/1506.
Frank, David John, Ann Hironaka, and Evan Schofer. "The Nation-State and the Natural Environment over the Twentieth Century." *American Sociological Review* 65, no. 1 (2000): 96–116.
Frederick, Jon. *Empty Promises: The IMF, the World Bank, and the Planned Failures of Global Capitalism*. Washington, DC: IMF/World Bank 50 Years Is Enough Network, 2003.
Friedman, Milton. *Capitalism and Freedom*. Chicago: University of Chicago Press, 1962.
Fukuyama, Francis. *The End of History and the Last Man*. New York: Free Press, 1992.
Giddens, Anthony. *Central Problems in Social Theory: Action, Structure, and Contradiction in Social Analysis*. Berkeley: University of California Press, 1979.
———. *The Constitution of Society: The Outline of the Theory of Structuration*. Cambridge: Polity, 1984.
Goldstein, B. D. "Toxicology and Risk Assessment in the Analysis and Management of Environmental Risk." In *Oxford Textbook of Public Health, Volume 2: The Methods of Public Health*, eds. R. Detels, R. Beaglehole, M. A. Lansing, and M. Gulliford. 931–939. Oxford: Oxford University Press, 2009.
Graham, Lorie M., and Siegfried Wiessner. "Indigenous Sovereignty, Culture, and International Human Rights Law." *South Atlantic Quarterly* 110, no. 2 (Spring 2011): 403–427.
Gramsci, Antonio. *Selections from the Prison Notebooks of Antonio Gramsci*, eds. Quintin Hoare and Geoffrey Nowell Smith. London: Lawrence and Wishart, 1971.
———. *Prison Notebooks*, Volumes 1–3, ed. J. A. Buttigieg. New York: Columbia University Press, 1996.
Greymorning, S. "Culture and Language: The Political Realities to Keep Trickster at Bay." *Canadian Journal of Native Studies* 20, no. 1 (2000): 181–196.
Griggs, Richard. "The Meaning of 'Nation' and 'State' in the Fourth World." In *CWIS Occasional Paper*. Olympia, WA: Center for World Indigenous Studies, 1992.
Gros, Daniel, et al. "The Case for IMF Quota Reform." *Council on Foreign Relations*. October 11, 2012. www.cfr.org/imf/case-imf-quota-reform.
Gudeman, Stephen. "Vital Energy: The Current of Relations." *Social Analysis* 56, no. 1 (2012): 57–73.

———. *The Anthropology of Economy: Community, Market, and Culture.* Malden, MA: Blackwell Publishers, 2001.

Guha, Ramachandra. *The Unquiet Woods: Ecological Change and Peasant Resistance in the Himalaya.* Berkeley: University of California Press, 2000.

Guha, Ramachandra, and Juan Martinez-Alier. *Varieties of Environmentalism.* London: Earthscan, 1997.

Haas, Peter M. "The Political Economy of Ecology: Prospects for Transforming the World Economy at Rio Plus 20." *Global Policy* 3, no. 1 (2012): 94–101.

Hall, Thomas, and James Fenelon. "The Futures of Indigenous Peoples: 9-11 and the Trajectory of Indigenous Survival and Resistance." *Journal of World Systems Research* 10, no. 1 (2004): 153–197.

Hansen, James, Makiko Sato, and Reto Ruedy. "Perception of Climate Change." *Proceedings of the National Academy of Sciences* 109, no. 37 (September 11, 2012): E2415–E2423.

Harmon, Amy, and Andrew Pollack. "Battle Brewing over Genetically Modified Food." *New York Times,* May 24, 2012.

Hayek, Friedrich von. *The Road to Serfdom.* Chicago: University of Chicago Press, 1944.

Hedges, Chris, and Joe Sacco. *Days of Destruction, Days of Revolt.* New York: Nation Books/Perseus Academic, 2012.

Hegel, Georg Wilhelm Friedrich. *Hegel's Philosophy of Right.* Translated by T. M. Knox. Neeland Media LLC, 1965 [1821].

Held, David. "Central Perspectives on the Modern State." *Political Theory and the Modern State,* ed. D. Held. 11–55. Cambridge: Polity Press, 1989.

Held, David, Anthony McGrew, David Goldblatt, and Jonathan Perraton. *Global Transformations: Politics, Economics and Culture.* Stanford: Stanford University Press, 1999.

Hine, R., J. N. Pretty, S. Twarog, and UUCBT Force. *Organic Agriculture and Food Security in Africa.* New York: United Nations, 2008.

Hipwell, William T. "A Deleuzian Critique of Resource-Use Management Politics in Industria." *The Canadian Geographer/Le Géographe Canadien* 48, no. 3 (2004): 356–377.

———. "The Industria Hypothesis: 'The Globalization of What?'" *Peace Review* 19, no. 3 (2007): 305–313.

Hobbes, Thomas. "Introduction." *Leviathan,* ed. C. B. Macpherson. New York: Penguin, 1985 [1651].

Holt-Giménez, Eric. "Out of AGRA: The Green Revolution Returns to Africa." *Development* 51 (2008): 464–471.

Hornberg, Alf, and Carole Crumley, eds. *The World System and the Earth System: Global Socioenvironmental Change and Sustainability since the Neolithic.* Walnut Creek, CA: Left Coast Press, 2007.

"Hottest Year on Record in the U.S. Might Still Be 2012, Despite Cooler October." *Huffington Post.* November 9, 2012. www.huffingtonpost.com2012/11/9/hottest-year-on-record-us-america-2012.

Howard, Bradley Reed. *Indigenous Peoples and the State: The Struggle for Native Rights.* Dekalb: Northern Illinois Press, 2003.

Huang, Chye-Ching, and Chad Stone. *Putting U.S. Corporate Taxes in Perspective.* Center on Budget and Policy Priorities. October 27, 2008. www.cbpp.org/cms?fa=view&=784.

"IDA Replenishment." *Bretton Woods Project: Critical Voices on the World Bank and IMF.* February 15, 2010. www.brettonwoodsproject.org/art-565921.

Illich, Ivan. *Energy and Equity.* San Francisco: Calder and Boyars, 1974.

Indigenous Peoples Global Conference on Rio+20 and Mother Earth. "Kari-Oca 2 Declaration." Rio de Janeiro, Brazil, 2012. www.ienearth.org/docs/DECLARATION-of-KARI-OCA-2-Eng.pdf.

International Indian Treaty Council/Consejo Internacional de Tratados Indios. "Declaration of Continuing Independence by the First International Indian Treaty Council." 1974.

International Indigenous Peoples Summit on Sustainable Development. "The Kimberley Declaration of the International Indigenous Peoples Summit on Sustainable Development," 2002. www.iwgia.org/sw217.asp.

International Monetary Fund. "IMF Quotas: Factsheet." www.imf.org/external/np/exr/facts/quotas.htm.

Jackson, Jeremy B. C. "Ecological Extinction and Evolution in the Brave New Ocean." *Proceedings of the National Academy of Sciences* 105, suppl. 1 (August 12, 2008): 11458–11465.

Jacques, Peter J. *Globalization and the World Ocean.* Lanham, MD: AltaMira/Rowman and Littlefield, 2006.

Jacques, Peter J., and Jessica Racine Jacques. "Monocropping Cultures into Ruin: The Loss of Food Varieties and Cultural Diversity." *Sustainability* 4, no. 10 (2012).

Jacques, Peter, Sharon Ridgeway, and Richard Witmer. "Federal Indian Law and Environmental Policy: A Social Continuity of Violence." *Journal of Environmental Law and Litigation* 18, no. 2 (2003).

James, Deborah. *Globalization: Leaving the WTO Behind.* Center for Economic and Policy Research. 2008. www.cepr.net/index2.php?option=com-content&taxk=view&id=1738&pop=1&page.

Jensen, H. G., and H. Zobbe. "Consequences of Reducing Limits on Aggregate Measurement of Support." In *Agricultural Trade Reform and the Doha Development Agenda,* eds. K. Anderson and W. Martin. 245–270. Washington, DC/Basingstoke, UK: World Bank/Palgrave Macmillan, 2006.

Kallis, Giorgos. "Droughts." *Annual Review of Environment and Resources* 33, no. 1 (2008): 85–118.

Kastner, T., M. J. I. Rivas, W. Koch, and S. Nonhebel. "Global Changes in Diets and the Consequences for Land Requirements for Food." *Proceedings of the National Academy of Sciences* 109 (2012): 6868–6872.

Kaur, G., R. S. Sawhney, and R. Vohra. "Micro-Cantilever Sensors for Detection of Pesticide Contents in the Water-Table of Malwa Region in Punjab." *International Journal of Advanced Research in Computer Engineering and Technology (IJARCET)* 1 (2012): 285–290.

Kaya, Ayse. "Conflicted Principals, Uncertain Agency: The International Monetary Fund and the Great Recession." *Global Policy* 3, no. 1 (2012): 24–34.

Keay, Douglas. "Interview for Woman's Own." In *Thatcher Archive.* London: Margaret Thatcher Foundation, 1987.

Kerr, R. B. "Lessons from the Old Green Revolution for the New: Social, Environmental and Nutritional Issues for Agricultural Change in Africa." *Progress in Development Studies* 12 (2012): 213–229.

Khor, Martin. "Agriculture: Falconer Report Says WTO Talks Failed on Political Factor."

Third World Network, August 13, 2008. www.ytnside.org.sg/title2/resurgence/215/cover4.doc.

———. "Trade: Rules Meeting Postponed, New Text Won't Be Issued." *Third World Network,* 2008. www.twnside.org.sg/title2/wto.info20080811.htm.

———. "WTO and the Third World: On a Catastrophic Course." In *Third World Traveler.* October/November 1999. www.thirdworldtraveler.com/WTO_MAI/.

Kirch, Patrick V. "Archaeology and Global Change: The Holocene Record." *Annual Review of Environment and Resources* 30, no. 1 (2005): 409–440.

Knutilla, Murray, and Wendee Kubik. *State Theories: Classical, Global and Feminist Perspectives.* Halifax, Nova Scotia: Fernwood Publishing, 2000.

Kormann, Carolyn. "Retreat of Andean Glaciers Foretells of Global Water Woes." April 9, 2009. http://e360.yale.edu.

Kütting, Gabriela. *Globalization and the Environment: Greening Global Political Economy.* Albany: State University of New York Press, 2004.

La Via Campesina. "The International Peasant's Voice." http://viacampesina.org/en/.

LaDuke, Winona, "Aspects of Traditional Knowledge and Worldview." In *Paradigm Wars: Indigenous Peoples' Resistance to Globalization,* eds. Jerry Mander and Victoria Tauli-Corpuz. 22–25. San Francisco: Sierra Club Books, 2006.

LaSalle, Tim, Paul Hepperly, and Amadou Diop. *The Organic Green Revolution.* Kutztown, PA: Rodale Institute, 2008.

Laxer, Gordon, and Dennis Soron, eds. *Not for Sale: Decommodifying Public Life.* Peterborough, Ontario: Broadview Press, 2006.

Leahy, Stephan. "Towards a New and Improved Green Revolution." Inter Press Service News Agency. 2008. www.ipsterraviva.net/Africa/print.asp?idnews=1791 6/18/.

———. "Indigenous Message to Rio+20: Leave Everything beneath Mother Earth." *National Geographic News Watch,* 2012.

Lee, K. "For Debate. The Impact of Globalization on Public Health: Implications for the UK Faculty of Public Health Medicine." *Journal of Public Health* 22, no. 3 (September 1, 2000): 253–262.

Leftwich, Adrien. "States of Underdevelopment: The Third World State in Perspective." *Journal of Theoretical Politics* 6, no. 1 (1994): 55–74.

Leopold, Aldo. *A Sand County Almanac, with Essays on Conservation from Round River.* New York: Oxford University Press/Ballantine Books Edition, 1970 [1949].

"Less Than Meets the Eye: IMF Reform Fails to Revolutionise the Institution." *Bretton Woods Project.* November 9, 2010. www.brettonwoodsproject.org/art-567128.

Levitus, S., et al. "World Ocean Heat Content and Thermosteric Sea Level Change (0–2000 m), 1955–2010." *Geophysical Research Letters* 39, no. 10 (2012): L10603.

Levitus, S., J. I. Antonov, T. P. Boyer, R. A. Locarnini, H. E. Garcia, and A. V. Mishonov. "Global Ocean Heat Content 1955–2008 in Light of Recently Revealed Instrumentation Problems." *Geophysical Research Letters* 36, no. 7 (2009): L07608.

Lewis, David Rich. "Native Americans and the Environment: A Survey of Twentieth-Century Issues." *American Indian Quarterly* 19, no. 3 (1995): 423–477.

Lipo, C. P., T. L. Hunt, and S. R. Haoa. "The 'Walking' Megalithic Statues (Moai) of Easter Island." *Journal of Archaeological Science,* 2012, in press corrected proof. 1–8.

Lo, Clarence Y. H. "Countermovements and Conservative Movements in the Contemporary U.S." *Annual Review of Sociology* 8 (1982): 107–134.

Looking Horse, Arvol. *White Buffalo Teachings from Chief Arvol Looking Horse, 19th*

Generation Keeper of the Sacred White Buffalo Pipe of the Lakota, Dakota and Nakota Great Sioux Nation. Williamsburg, MA: Dreamkeepers Press, 2001.
Macy, Joanna. *World as Lover, World as Self.* San Francisco: Paralax Press, 1991.
Maffi, Luisa. "Bio-Cultural Diversity for Endogenous Development: Lessons from Research, Policy, and On-the-Ground Experiences." Paper presented at the conference Endogenous Development and Bio-Cultural Diversity: The Interplay of Worldviews Globalization and Locality, Geneva, Switzerland, October 3–5, 2006.
———. "Endangered Languages, Endangered Knowledge." *International Social Science Journal* 54, no. 173 (2002): 385–393.
———. "Linguistic, Cultural, and Biological Diversity." *Annual Review of Anthropology* 34 (2005): 599–617.
Mander, Jerry. "Foreword: The Resistance to Southern Perspectives." In *Views from the South: The Effects of Globalization and the WTO on Third World Countries,* ed. Sarah Anderson. 1–6. Chicago: Food First Books and the International Forum on Globalization, 2000.
Manjunatha, B., D. Hajong, and R. Koulagi. "Impact of Agricultural Development on Environment." *Environment and Ecology* 30 (2012): 393–396.
Marshall, Peter. *Demanding the Impossible: A History of Anarchism.* Oakland, CA: PM Press, 2010.
Marten, Gerry. "'Non-Pesticide Management' for Agricultural Pests: Escaping the Pesticide Trap—Controlling Pests without Chemical Pesticides Catapults Farmers from Chronic Poisoning and Debt to Health and Hope." *The Eco-Tipping Points Project,* 2012. www.ecotippingpoints.org/our-stories/indepth/india-pest-management-nonpesticide-neem.html.
Martínez-Torres, María Elena, and Peter M. Rosset. "La Vía Campesina: The Birth and Evolution of a Transnational Social Movement." *Journal of Peasant Studies* 37, no. 1 (January 1, 2010): 149–175.
Marx, Karl, ed. *The Eighteenth Brumaire of Louis Bonaparte,* ed. Friedrich Engels. 2006 [1852]. www.marxists.org/archive/marx/works/1852/18th-brumaire/.
Maybury-Lewis, David. *Indigenous Peoples, Ethnic Groups, and the State.* Boston: Allyn and Bacon, 1996.
Mayet, Mariam. "The New Green Revolution in Africa: A Trojan Horse for GMOs." In *Voices from Africa: African Farmers and Environmentalists Speak Out against a New Green Revolution in Africa,* eds. Anuradha Mittal and Melissa Moore. 2–7. Oakland, CA: The Oakland Institute, 2009.
McGregor, Deborah. "Coming Full Circle: Indigenous Knowledge, Environment, and Our Future." *American Indian Quarterly* 28, no. 3/4 (2004): 385–410.
Meadows, Donnella, Jorgen Randers, and Dennis Meadows. *Limits to Growth: The 30-Year Update.* White River Junction, VT: Chelsea Green Publishing, 2004.
Mies, Maria, and Veronika Bennholdt-Thomsen. *The Subsistence Perspective: Beyond the Globalised Economy.* New York: Zed Books, 1999.
Milanovic, Branko. "The Two Faces of Globalization: Against Globalization as We Know It." *World Development* 31, no. 4 (2003): 667–683.
Miles, Tom. "WTO Sees Tentative Progress in Trade Talks after Stalemate." Reuters, October 13, 2012. www.reuters.com/assets/print?aid.
Miliband, Ralph. *The State in Capitalist Society.* New York: Basic Books, 1969.
Millennium Ecosystem Assessment. *Ecosystems and Human Well-Being: Our Human Planet: Summary for Decision-Makers.* Washington, DC: Island Press, 2005.

Mohanty, B. B. "'We Are Like the Living Dead': Farmer Suicides in Maharashtra, Western India." *Journal of Peasant Studies* 32, no. 2 (2005): 243–276.

Mokhiber, Russell, and Robert Weissman. "Corporate Globalization and the Poor." *New Renaissance Magazine*. www.ru.org/econotes/corporate-globalization-and-the-poor.

Montejo, Victor. "Mayan Ways of Knowing: Modern Mayans and the Elders." In *The Will to Survive: Indigenous Essays on the Politics of Culture, Language, and Identity*, ed. Stephen Greymorning. 154–170. New York: McGraw-Hill, 2004.

Mooney, Chris. "The Truth about Fracking." *Scientific American* 305, no. 5 (2011): 80–85.

Moore, Kathleen Dean, and Michael P. Nelson. *Moral Ground: Ethical Action for a Planet in Peril*. San Antonio, TX: Trinity University Press, 2010.

Mushita, Andrew, and Carol Thompson. *Biopiracy of Biodiversity: Global Exchange as Enclosure*. Trenton, NJ: Africa World Press, 2007.

Myers, Samuel S., and Jonathan A. Patz. "Emerging Threats to Human Health from Global Environmental Change." *Annual Review of Environment and Resources* 34, no. 1 (2009): 223–252.

Nash, June C. *Mayan Visions: The Quest for Autonomy in an Age of Globalization*. New York: Routledge, 2001.

Naylor, Rosamond L., and Walter P. Falcon. "Food Security in an Era of Economic Volatility." *Population and Development Review* 36, no. 4 (2010): 693–723.

Nicholls, R. J., et al. "Coastal Systems and Low-Lying Areas." In *Climate Change 2007: Impacts, Adaptation and Vulnerability. Contribution of Working Group II to the Fourth Assessment Report of the Intergovernmental Panel on Climate Change*, eds. M. L. Parry et al. 315–356. Cambridge: Cambridge University Press, 2007.

Oak, Robert. "House Passed Bill Which Closes the 'Offshore Outsourcing' International Corporate Tax Scheme." *Economic Populist*, 2010.

O'Brien, Karen, et al. "Mapping Vulnerability to Multiple Stressors: Climate Change and Globalization in India." *Global Environmental Change* 14 (2004): 303–313.

O'Brien, Sharon. *American Indian Tribal Governments*. The Civilization of the American Indian Series. Norman: University of Oklahoma Press, 1989.

Okogbule, Nlerum S. "Globalization, Economic Sovereignty, and African Development: From Principles to Realities." *Journal of the Third World Studies* (Spring 2008). http://find articles.com/p/articlesmi_qa3821/is_200806/ai_n25418590/print. 6/18/2008.

Oland, M., S. M. Hart, and L. Frink, eds. *Decolonizing Indigenous Histories: Exploring Prehistoric/Colonial Transitions in Archaeology*. Tuscon, AZ: University of Arizona Press.

Organski, Abramo Fimo Kenneth. *The Stages of Political Development*. New York: Knopf, 1965.

Ostler, Jeffrey. "'They Regard Their Passing as Wakan': Interpreting Western Sioux Explanations for the Bison's Decline." *The Western Historical Quarterly* 30, no. 4 (1999): 475–497.

Paarlberg, Robert. *Food Politics: What Everyone Needs to Know*. New York: Oxford University Press, 2010.

Paehlke, Robert. *Democracy's Dilemma: Environment, Social Equity, and the Global Economy*. Cambridge, MA: MIT Press, 2004.

Parkinson, Claire L. "Earth's Cryosphere: Current State and Recent Changes." *Annual Review of Environment and Resources* 31, no. 1 (2006): 33–60.

Parry, M. L., O. F. Canziani, J. P. Palutikof, P. J. van der Linden, and C. E. Hanson, eds. *Climate Change 2007: Impacts, Adaptation and Vulnerability. Contribution of Working Group II to the Fourth Assessment Report of the Intergovernmental Panel on Climate Change.* Cambridge: Cambridge University Press, 2007.

Patel, Raj. *Stuffed and Starved: Markets, Power and the Hidden Battle for the World Food System.* London: Portobello Books, 2007.

Paterson, Matthew. *Understanding Global Environmental Politics: Domination, Accumulation and Resistance.* New York: St. Martin's Press, 2000.

Pearce, Fred. "Arctic Sea Ice May Have Passed Crucial Tipping Point." *New Scientist,* no. 2858 (2012).

Peck, Jamie, and Adam Tickell. "Apparitions of Neoliberalism: Revisiting 'Jungle Law Breaks Out.'" *Area* 44, no. 2 (2012): 245–249.

———. "Neoliberalizing Space." *Antipode* 34, no. 3 (2002): 380–404.

Peet, Richard. *Unholy Trinity: The IMF, World Bank and WTO.* 2nd ed. New York: Zed Books, 2009.

Peltier, Leonard. *Prison Writings: My Life Is My Sun Dance.* New York: St. Martin's Griffin, 1999.

Peters, G. P., G. Marland, C. Le Quéré, T. Boden, J. G. Canadell, and M. R. Raupach. "Rapid Growth in CO_2 Emissions after the 2008–2009 Global Financial Crisis." *Nature Climate Change,* 2011.

Petras, James, and Henry Veltmeyer. *Globalization Unmasked: Imperialism in the 21st Century.* New York: Zed Books, 2003.

Pimentel, David, Paul Hepperly, Rita Seidel, James Hanson, and David Douds. "Environmental, Energetic, and Economic Comparisons of Organic and Conventional Farming Systems." *Bioscience* 55, no. 7 (2005): 573–582.

Pingali, P. L. "Green Revolution: Impacts, Limits, and the Path Ahead." *Proceedings of the National Academy of Sciences* 109 (2012): 12302–12308.

Polak, Paul. *Out of Poverty.* San Francisco: Berret-Koehler, 2008.

Poulantzas, Nicos. "The Problem of the Capitalist State." In *Ideology and Social Science,* ed. R. Blackburn. 238–253. London: Fontana, 1972.

Presbitero, Andrea F., and Alberto Zazzaro. "IMF Lending in Times of Crisis: Political Influences and Crisis Prevention." *World Development* 40, no. 10 (2012): 1944–1969.

Pretty, J. "Agricultural Sustainability: Concepts, Principles and Evidence." *Philosophical Transactions of the Royal Society B: Biological Sciences* 363, no. 1491 (2008): 447–465.

Princen, Thomas. *The Shadows of Consumption: Consequences for the Global Environment.* Cambridge, MA: MIT Press, 2008.

Princen, Thomas, Michael Maniates, and Ken Conca. *Confronting Consumption.* Cambridge, MA: MIT Press, 2002.

Prucha, Francis Paul. *The Great Father: The United States Government and the American Indians.* Lincoln: University of Nebraska Press, 1984.

Pyle, George. *Raising Less Corn, More Hell: The Case for the Independent Farm and against Industrial Food.* New York: Public Affairs/Perseus Books Group, 2005.

Raghavan, Chakravarthi. "Seattle WTO Ministerial Ends in Failure." *Third World Network,* 1999, www.twnside.org.sg/title/deb2-cn.htm.

Rand, Ayn. *The Fountainhead.* New York: Bobbs-Merrill Company, 1971.

Redford, Kent H., and J. Peter Brosius. "Diversity and Homogenization in the Endgame." *Global Environmental Change* 16, no. 4 (2006): 317–319.

Reinhardt, Akim D. *Ruling Pine Ridge: Oglala Lakota Politics from the Ira to Wounded Knee.* Lubbock: Texas Tech University Press, 2007.

Robbins, Richard. *Global Problems and the Culture of Capitalism*. Upper Saddle River, NJ: Prentice-Hall, 2010.

———. "The Guarani: The Economics of Ethnocide." In *The Indigenous Experience: Global Perspectives*, eds. Roger Maaka and Chris Andersen. 150–157. Toronto: Canadian Scholar's Press, 2006.

Roberts, Michael J., Wolfram Schlenker, and Jonathan Eyer. "Agronomic Weather Measures in Econometric Models of Crop Yield with Implications for Climate Change." *American Journal of Agricultural Economics*, May 22, 2012: aas047.

Robertson, Robbie. *The Three Waves of Globalization: A History of a Developing Global Consciousness*. New York: Zed Books, 2003.

Robinson, William I. "Global Capitalism Theory and the Emergence of Transnational Elites." *Critical Sociology* 38, no. 3 (2012): 349–363.

Robinson, William I., and Jerry Harris. "Towards a Global Ruling Class? Globalization and the Transnational Capitalist Class." *Science and Society* 64, no. 1 (2000): 11–54.

Rockstrom, Johan, et al. "Planetary Boundaries: Exploring the Safe Operating Space for Humanity." *Ecology and Society* 14, no. 2 (2009). www.ecologyandsociety.org/vol14/iss2/art32/.

———. "A Safe Operating Space for Humanity." *Nature* 461, no. 7263 (2009): 472–475.

Rodney, Walter. *How Europe Underdeveloped Africa*. Washington, DC: Howard University Press, 1982.

Romaine, Suzanne. "Planning for the Survival of Linguistic Diversity." *Language Policy* 5, no. 4 (2006): 443–475.

Rosset, Peter. "Food Sovereignty and the Contemporary Food Crisis." *Development* 51, no. 4 (2008): 460–463.

———. "The Time Has Come for La Via Campesina and Food Sovereignty." *Focus on the Global South*, 2008.

Rostow, Walt W. *Stages of Economic Growth: A Non-Communist Manifesto*. Cambridge: Cambridge University Press, 1960.

Rousseau, Jean-Jacques. *The Social Contract*. New York: Penguin, 1968 [1762].

Sánchez, P. A., M. S. Swaminathan, P. Dobie, and N. Yuksel. *Halving Hunger: It Can Be Done*. London/Sterling VA: United Nations Millennium Project Task Force on Hunger, 2005.

Santos, Boaventura De Sousa. "The World Social Forum and the Global Left." *Politics and Society* 36, no. 2 (June 1, 2008): 247–270.

Saragih, Henry. "Food Sovereignty to Answer World Food and Energy Crisis." La Via Campesina, www.viacampesina.org/en/index.php?option=com_content&view=article&id=566:food-sovereignty-to-answer-world-food-and-energy-crisis&catid=22:biodiversity-and-genetic-resources&Itemid=37.

Sarkar, A. "Sustaining Livelihoods in Face of Groundwater Depletion: A Case Study of Punjab, India." *Environment, Development, and Sustainability* 14 (2012): 183–195.

Schäfer, Mike S. "Online Communication on Climate Change and Climate Politics: A Literature Review." *Wiley Interdisciplinary Reviews: Climate Change*, 2012.

Schlenker, Wolfram, and Michael J. Roberts. "Nonlinear Temperature Effects Indicate Severe Damages to U.S. Crop Yields under Climate Change." *Proceedings of the National Academy of Sciences (PNAS)* 106, no. 37 (September 15, 2009): 15594–15598.

Scott, James C. *The Moral Economy of the Peasant: Rebellion and Subsistence in Southeast Asia*. New Haven: Yale University Press, 1977.

———. *The Art of Not Being Governed: An Anarchist History of Upland Southeast Asia.* New Haven: Yale University Press, 2009.
———. *Two Cheers for Anarchism: Six Easy Pieces on Autonomy, Dignity, and Meaningful Work and Play.* Princeton: Princeton University Press, 2012.
Sen, Amartya Kumar. *Poverty and Famines: An Essay on Entitlement and Deprivation.* New York: Oxford University Press, 1981.
Shandra, Carrie L., John M. Shandra, and Bruce London. "The International Monetary Fund, Structural Adjustment, and Infant Mortality: A Cross-National Analysis of Sub-Saharan Africa." *Journal of Poverty* 16, no. 2 (April 1, 2012): 194–219.
Shiva, Vandana. *Earth Democracy: Justice, Sustainability, and Peace.* Cambridge, MA: South End Press. 2005.
———. "Globalization and Poverty." In *There Is an Alternative: Subsistence and Worldwide Resistance to Corporate Globalization,* eds. Veronika Bennholdt-Thomsen, Nicholas Faraclas, and Claudia von Werlhof. 57–66. New York: Zed Books, 2001.
———. *The Violence of the Green Revolution: Third World Agriculture, Ecology and Politics.* 4th ed. New York: Zed Books, 2000.
Shultz, Jim. "The Cochabamba Water Revolt and Its Aftermath." In *Dignity and Defiance: Stories from Bolivia's Challenge to Globalization,* eds. Jim Shultz and Melissa Crane Draper. 9–44. Berkeley: University of California Press, 2008.
Sklair, Leslie. *Globalization: Capitalism and Its Alternatives.* 3rd ed. New York: Oxford University Press, 2002.
———. "Transnational Capitalist Class." In *The Wiley-Blackwell Encyclopedia of Globalization,* ed. George Ritzer. Hoboken, NJ: Blackwell Publishing, 2012.
Skogstad, G. "Internationalization, Democracy, and Food Safety Measures: The (Il)Legitimacy of Consumer Preferences." *Global Governance* 7 (2001): 293.
Smith, Jackie, and Dawn Wiest. *Social Movements in the World-System: The Politics of Crisis and Transformation.* American Sociological Association's Rose Series in Sociology. New York: Russell Sage Foundation, 2012.
Southgate, Douglas, Douglas Graham, and Luther Tweeten. *The World Food Economy.* Malden, MA: Wiley-Blackwell, 2007.
Srivastava, P., M. Gupta, and S. Mukherjee. "Mapping Spatial Distribution of Pollutants in Groundwater of a Tropical Area of India Using Remote Sensing and GIS." *Applied Geomatics* 4 (2012): 21–32.
Starr, Amory. *Naming the Enemy: Anti-Corporate Movements Confront Globalization.* London: Pluto Press, 2000.
Steffen, Will, Johan Rockström, and Robert Costanza. "How Defining Planetary Boundaries Can Transform Our Approach to Growth." *Solutions* 2, no. 3 (2011). http://thesolutionsjournal.anu.edu.au/node/935.
Steinki, Bernhard, and Wolfgang Bergthaler. "Recent Reforms of the Finances of the International Monetary Fund: An Overview." In *European Yearbook of International Economic Law* (Eyiel), eds. Christoph Herrmann and Jörg Philipp Terhechte. 635–666. Berlin: Springer, 2012.
Stewart-Harawira, Makere. *The New Imperial Order: Indigenous Responses to Globalization.* New York: Zed Books, 2005.
———. "Returning the Sacred: Indigenous Ontologies in Perilous Times." In *Radical Human Ecology: Intercultural and Indigenous Approaches,* eds. Lewis Williams, Rose Roberts, and Alastair McIntosh. 73–88. Farnham, UK: Ashgate, 2012.
Structural Adjustment Participatory Review International Network (SAPRIN). *Structural*

Adjustment: The SAPRI Report, the Policy Roots of Economic Crisis, Poverty and Inequality. New York: Zed Books, 2004.

Stuart, Elizabeth. *Blind Spot: The Continued Failure of the World Bank and IMF to Fully Assess the Impact of Their Advice on Poor People.* Oxford: Oxfam International, 2007.

Sutherland, W. J. "Parallel Extinction Risk and Global Distribution of Languages and Species." *Nature* 423, no. 6937 (2003): 276–279.

Szasz, Andrew. *Shopping Our Way to Safety: How We Changed from Protecting the Environment to Protecting Ourselves.* Minneapolis: University of Minnesota Press, 2007.

Tainter, Joseph A. *The Collapse of Complex Societies.* Cambridge: Cambridge University Press, 1988.

Tangel, Andrew, and Alejandro Lazo. "JP Morgan Targeted by U.S. Task Force for Role in Financial Mess." *Los Angeles Times,* October 2, 2012.

Taylor, M. Scott. "Buffalo Hunt International Trade and the Virtual Extinction of the North American Bison." *American Economic Review* 101, no. 7 (2006): 3162–3195.

Terena, M. Marcos. "Native Peoples and the World Social Forum." *Inter Press Service News Agency,* 2008. www.forumsocialmundial.org.br/noticias_textos.php?cd_news=443.

Thompson, Carol B. "Alliance for a Green Revolution in Africa (AGRA): Advancing the Theft of African Genetic Wealth." *Review of African Political Economy* 39 (2012): 345–350.

Tinker, George. "An American Indian Theological Response to Eco-Justice." In *Defending Mother Earth: Native American Perspectives on Environmental Justice,* ed. Jace Weaver. Maryknoll, NY: Orbis Books, 1996.

Tippett, Krista. "Joanna Macy: A Wild Love for the World." American Public Media Programs, September 16, 2010. http://being.publicradio.org/programs/2010/wild-love-for-world/transcript.shtml.

Townsend, Patricia K. *Environmental Anthropology: From Pigs to Policies.* 2nd ed. Groveland, IL: Waveland Press, 2009.

Turner, B. L., and Paul Robbins. "Land-Change Science and Political Ecology: Similarities, Differences, and Implications for Sustainability Science." *Annual Review of Environment and Resources* 33, no. 1 (2008): 295–316.

Turner, Graham M. "A Comparison of the Limits to Growth with 30 Years of Reality." *Global Environmental Change* 18, no. 3 (2008): 397–411.

United Nations Food and Agriculture Organization. "FAO Initiative on Soaring Food Prices." www.fao.org/fileadmin/user_upload/ISFP/pdf_for_site_Country_Response_to_the_Food_Security.pdf.

United Nations Secretariat of the Permanent Forum on Indigenous Issues. *State of the World's Indigenous Peoples.* New York: United Nations Publications, 2009.

"United States Data." World Bank. 2012. http//data.worldbank.org/country/united-states.

"U.S. on Track for Warmest Year on Records, and Second Most Extreme." ThinkProgress. November 12, 2012. http://thinkprogress.org/climate/2012/11/12/1177671/us-on-track-for-warmest-year-on-record.

Venne, Sharon. "She Must Be Civilized: She Paints Her Toe Nails." In *A Will to Survive: Indigenous Essays on the Politics of Culture, Language, and Identity,* ed. Stephen Greymorning. New York: McGraw-Hill, 2004.

Vidal, John. "UN Warns of Looming Worldwide Food Crisis in 2013." *The Guardian.* October 13, 2012. www.guardian/co.uk/global-development/2012/oct/14/un-warns.

Virmani, Arvind. "Quota Formula Reform Is about IMF Credibility." *VOX*, June 22, 2012. www.voxeu.org/article/quota-formula-reform-about-imf-credibility.
Vitousek, Peter M., Paul R. Ehrlich, Anne H. Ehrlich, and Pamela A. Matson. "Human Appropriation of the Products of Photosynthesis." *Bioscience* 36, no. 6 (1986): 368–373.
Vitousek, Peter M., Harold A. Mooney, Jane Lubchenco, and Jerry Melillo. "Human Domination of Earth's Ecosystems." *Science* 277 (July 15, 1997): 494–499.
Vogli, Roberto De. *Progress or Collapse: The Crises of Market Greed*. New York: Routledge, 2013.
Voices from Africa: African Farmers and Environmentalists Speak Out against a New Green Revolution in Africa. Oakland, CA: The Oakland Institute, 2009. http://media.oaklandinstitute.org/sites/oaklandinstitute.org/files/voicesfromafrica_full.pdf.
Vreeland, James Raymond. *The IMF and Economic Development*. Cambridge: Cambridge University Press, 2003.
Wackernagel, Mathis, et al. "Tracking the Ecological Overshoot of the Human Economy." *Proceedings of the National Academy of Science (PNAS)* 99, no. 14 (2002): 9266–9271.
Wade, Robert Hunter. "What Strategies Are Viable for Developing Countries Today? The World Trade Organization and the Shrinking of 'Development Space.'" *Review of International Political Economy* 10, no. 4 (2003): 621–644.
———. "U.S. Keeps Control of the World Bank." *Le Monde Diplomatique*, English ed. October 24, 2012. http://mondediplo.com/blogs/us-keeps-control-of-world-bank.
Wall, Dennis, and Virgil Masayesva. "People of the Corn: Teachings in Hopi Traditional Agriculture, Spirituality, and Sustainability." *American Indian Quarterly* 28, no. 3/4 (2004): 435–454.
Wallach, Lori. "RIP Doha Round." *Public Citizen*, 2008.
Wallach, Lori, and Patrick Woodall. *Whose Trade Organization? A Comprehensive Guide to the WTO*. New York: The New Press, 2004.
Wallerstein, Immanuel. *World-Systems Analysis: An Introduction*. Durham, NC: Duke University Press, 2004.
———. *The Modern World-System I: Capitalist Agriculture and the Origins of the European World-Economy in the Sixteenth Century*. Berkeley: University of California Press, 2011.
Ward, Graeme, and Claire Smith. *Indigenous Cultures in an Interconnected World*. London: Allen and Unwin, 2000.
Waters, Frank. *The Book of Hopi: The First Revelation of the Hopi's Historical and Religious World-View of Life*. New York: Penguin Books, 1963, repr. 1971.
Watson, Reg A., et al. "Global Marine Yield Halved as Fishing Intensity Redoubles." *Fish and Fisheries*, 2012.
Wei, Ting, et al. "Developed and Developing World Responsibilities for Historical Climate Change and CO_2 Mitigation." *Proceedings of the National Academy of Sciences* 109, no. 32 (August 7, 2012): 12911–12915.
Weisbrot, Mark, Robert Naiman, and Joyce Kim. *The Emperor Has No Growth: Declining Economic Growth Rates in the Era of Globalization*. Center for Economic and Policy Research, 2000.
Weiss, Gary. *Ayn Rand Nation: The Hidden Struggle for America's Soul*. New York: St. Martin's Press, 2012.
Wilmer, Franke. *The Indigenous Voice in World Politics: Since Time Immemorial*. Thousand Oaks, CA: Sage Publications, 1993.
Wildcat, Daniel R. "Technological Homelessness." In *Power and Place: Indian Education*

in America, eds. Vine Deloria Jr. and Daniel R. Wildcat. 67–78. Golden, CO: Fulcrum Resources, 2001.

Wilson, Angela Cavender. "Reclaiming Our Humanity: Decolonization and the Recovery of Indigenous Knowledge." In *War and Border Crossings: Ethics When Cultures Clash,* eds. Peter A. French and Jason A. Short. 255–264. Lanham, MD: Rowman and Littlefield, 2005.

Winton, M. "Does the Arctic Sea Ice Have a Tipping Point?" *Geophysical Research Letters* 33 (2006): L23504.

Wise, Timothy A. "Promise or Pitfall? The Limited Gains from Agricultural Trade Liberalisation for Developing Countries." *Journal of Peasant Studies* 36, no. 4 (October 1, 2009): 855–870.

Woodward, David. *The IMF and World Bank in the 21st Century: The Need for Change.* London: New Economic Foundation, 2005.

World Bank and IMF Emergency Loans: A Cure or Cause of the Food Crisis? Brussels, Belgium: European Network on Debt and Development, 2008.

World Bank. "United States Data." World Bank, 2012. http://data.worldbank.org/country/united-states.

World Conference of Indigenous Peoples on Territory, Environment and Development. "Indigenous Peoples Earth Charter and Kari-Oca Declaration." Kari Oca, Brazil, 1992. www.trc.org.nz/sites/trc.org.nz/files/Indigenous%20Peoples%20Earth%20Charter.pdf.

World People's Conference on Climate Change and the Rights of Mother Earth. "The People's Agreement." April 24, 2010. www.pwccc.wordpress.com/2010/04/24/peoples-agreement/.

INDEX

Agriculture: African, 87–90; enclosure of the commons, 56–57; industrial, 5, 21, 30, 34, 48, 50, 65, 67, 81–83, 85, 89, 140; intercropping, 84; Lakota and, 135; monocropping, 48, 182, 186, 151; the state and, 98–99; subsidies, 59–63, 86–87; subsistence/traditional, 13, 38, 73, 79, 84, 87–89. *See also* World Trade Organization
Alliance for a Green Revolution in Africa (AGRA), 87–90
American Indian Movement (AIM), 135–136, 141
Anasazi, 132
Anglo countries. *See* Australia; Canada; New Zealand; United Kingdom; United States
Argentina, 37
Austerity measures, 45, 119, 124
Australia, 2, 60, 63, 73, 86, 92

Bakunin, Michael, 94
Bali Indigenous Peoples Political Declaration, 143
Barry, John, 106–107
Bechtel Corporation, 66–67
Bello, Walden, 6
Bengal famine. *See* Hunger and malnutrition
Big Foot (Lakota), 135
Bill and Melinda Gates Foundation. *See* Alliance for a Green Revolution in Africa
Biodiversity, 5–6, 12, 56–59, 71, 121, 140, 148
Biopiracy, 58, 88, 138
Biotechnology, 58, 87–90. *See also* Patents; Food

Bison, nation of, 135–138
Bookchin, Murray, 107, 115
Borloug, Norman, 81. *See also* Food, industrialization of (Green Revolution)
Botswana, 87
Brazil, 3, 5, 29, 63, 74, 91, 121, 142
Bretton Woods Institutions. *See* International Monetary Fund; World Bank; World Trade Organization
Buen vivir, 4, 75, 78, 118, 121–122, 125–127, 144, 153
Bunker, Stephen, 69, 110–112
Burke, Edmund, 122
Bush, George H. W., 24, 70
Bush, George W., 4, 17, 25, 55, 89, 124

Canada, 7, 44, 60, 86. 92, 135
Cancer, 20, 54, 83
C'anupa Bundle, the Sacred. *See* Looking Horse, Chief Arvol
Carbon dioxide (CO_2), 5, 69
Caterpillar Corporation, 86
Charter of Indigenous Peoples of the Arctic and the Far East Siberia, 143
Charter of the International Alliance of Indigenous and Tribal Peoples of the Tropical Forests, 143
Chevron Corporation, 86
China, greenhouse gas emissions and, 3
Chlorofluorocarbons (CFCs), 20
Civilization: crisis of, 74; *industria* and, 17, 96, 145; modern/Western, 7, 11, 102, 120; new/adaptive, 15, 152–153; non-Western, 108; trade/development and, 24, 34

Climate change, viii, 2–3, 5–6, 13, 21, 27, 63–34, 75, 90, 119–120, 132, 143–144, 149
Cochabamba, 53; Declaration, 4–6
Codex. *See* World Trade Organization
Columbia, 33
Comte, Auguste, 100–101
Core. *See* World capitalist system
Crazy Horse, 134. *See also* Lakota and Nakota
Cultures and biodiversity (bioculture) parallel extinction risk, 5, 8
Custer, General George Armstrong, 135

Dead zones, 140
Declaration of the Indigenous Peoples of Eastern Africa in the Regional WSSD Preparatory Meeting, 143
Deleuze, Gilles, 17–19, 74, 96
Deloria, Vine, 1, 129
Democracy Center, 53
Development: discourse, 21, 30–35; economic, 21; international financial institutions and, 25; theory, 93; underdevelopment, 30
Division of labor, 14–16, 99, 104, 115
Doctrine of discovery, 102
Drought, 2–3, 48–49, 59–60, 64, 87, 90, 132
Dryzek, John, 106
Durkheim, Emile, 101, 104

Eckersley, Robyn, 106
Ecological unequal exchange, 110–111, 113
Economic Structural Adjustment Program (ESAPs), 87. *See also* International Monetary Fund (IMF), structural adjustment programs
Engineers without Borders/Ingenieurs sans Frontieres, 151
Escobar, Arturo, 30
Europe: colonization, 11; European Union, 28, 54, 86; mind/mode, 11; monetary zone/Eurozone, 28–29
Federal Bureau of Investigation (FBI), 136
Food: biodiversity and, 5, 87, 89, 90; cash crops, 48, 72, 80, 87; climate and, 63–64, corporations and, 85, 89, 94; crisis, 48, 64, 79, 87; distribution and transportation, 85; genetically modified, 21, 89; importation and exportation, 64, 80–81, 87, 89; industrialization of (Green Revolution), 34, 71, 81–83, 86–88, 90, 93, 108, 140, 147, 149, 151; organic movement, 149; pollution from, 140; prices, 64, 79–81, 85–86, 119; production and yield, 48, 57, 64, 88; riots, 38, 79; security, 47–48, 50, 57, 63, 80, 87, 89, 119; sovereignty, 5, 94; standards and safety, 54; subsistence, 56, 84, 136; supply system, 48, 59–61; sustainability and, 120–121; webs, 2
Ford Foundation, 33, 81, 93
Fourth World, 9–10, 98, 132–133
France, 28, 44

G20, 29. *See also specific countries*
General Agreement on Tariffs and Trade (GATT), 21, 24, 36, 49–50, 55, 59, 61; Uruguay Round, Ninth, 49, 56, 59–60, 62, 80. *See also* World Trade Organization
Genetically modified organisms (GMOs), 21
Geoculture, 15–16, 46
Germany, 28, 44
Giddens, Anthony, 105, 115
Globalization: anti-, 73; corporate and economic, 5–9, 13–14, 19–20, 24, 30, 32, 36, 40, 49–51, 53, 62, 65–68, 70, 72–73, 75, 77–78, 80, 92, 98, 109, 116, 147, 153
Global warming, 2, 5; oceans, 2
Goegescu-Roegen, Nicholas, 111
Goldman, Emma, 94
Gramsci, Antonio, 100, 105, 115
Great Irish Potato Famine (1845–1849), 84. *See also* Hunger and malnutrition
Great Recession (2008), 28–29, 44–45, 121, 123–124
Greece, 124
Green Revolution. *See* Food, industrialization of
Greenspan, Alan, 124
Guatemala, 132
Guattari, Feliz, 17, 96

Hayek, Friedrich von, 93
Hedge funds, 124
Hegel, Georg, 101
Hipwell, William, 14, 17–20, 24, 46, 74, 95–96
Hobbes, Thomas, 97, 102
Hopi, 9, 12, 97–98, 132–134
Hunger and malnutrition, 64, 80, 82, 87, 127; Bengal famine of 1943, 79; famine, 61, 87, 90, 136; Great Irish Potato Famine (1845–1849), 84
Hydrologic societies. *See* Worster, Donald

Identitarian politics, 17–20, 22, 30–31, 49, 126
Idle No More movement, 152
Illich, Ivan, 151
India, 3, 57–58, 62–63, 73–74, 93; Chipko Movement, 149; Punjab, 81–83
Indian Ocean, 2
Indigenous People's Earth Charter, 121, 142–143, 143

Indigenous Women's Network of Latin America and the Caribbean for Biodiversity (RMIB), 4
Indonesia, 142
International Bank for Reconstruction and Development (IBRD). *See* World Bank
International Crop Research Institute for the Semi-Tropics, 88
International Development Association (IDA). *See* World Bank
International Finance Corporation (IFC). *See* World Bank
International Indian Treaty Council (IITC), 141
International Monetary Fund (IMF), 7, 11, 21, 23–29, 32, 35–41, 44–45, 47–49, 52–53, 60, 80, 87; structural adjustment programs (SAPs), 37–38, 41–42, 44, 47–48, 52, 60, 87–88
Italy, 44

Japan, 28, 39, 44, 60, 69, 108, 150
John Hancock Life Insurance Company, 86
JP Morgan Chase, 123–124

Kachinas, 134. *See also* Hopi
Kari-Oca I, 4, 142–144
Kari-Oca II, 4, 121, 144
Keynes/Keynesian economics, 44, 93
Khoi-san peoples, 143
Kimberley Declaration, 143. *See also* World Summit on Sustainable Development
Kropotkin, Peter, 94

Lakota and Nakota, 134–137; Guardians of the Oglala Nation (GOON Squad), 135; *Mitakouye oyasin* ("All my relations"), 137; Oglala, 135; Pine Ridge Reservation, 135; Standing Rock Reservation, 135; White Buffalo Calf Woman, 136–137; Wounded Knee Massacre (1876), 135; Wounded Knee Siege (1973), 135, 141
Leticia Declaration of Indigenous Peoples and Other Forest Dependent Peoples on the Sustainable Use and Management of All Types of Forests, 143
Little Bighorn, Battle of (1876), 135. *See also* Lakota and Nakota
Looking Horse, Chief Arvol, 137, 153
Luxemburg, Rosa, 67, 70–72

Malaysia, 62
Maori, 130–131, 138. *See also* New Zealand
Marshall Plan, 32
Marx, Karl, 71, 76, 101
Marxism, 100, 116; instrumentalism, 103; structuralism, 105; theories of imperialism, 103; theory of the state, 102–104
Mataatua Declaration, 143
Maya, 77–78, 84, 128, 131–132, 138
Maybury-Lewis, David, 97
Means, Russell, 135
Mexico, 4, 36, 41, 43, 45, 55, 81, 86, 132, 140. *See also* Zapatistas
Mogollan peoples. *See* Pueblo peoples
Monarchies, absolute, 99, 108, 110
More and Better Network, 73

Neoliberalism, 11, 15–16, 19, 24, 35–40, 45, 74, 93–94, 119, 145
New world order, 24, 29, 38, 45
New Zealand, 130
Nitrates, phosphorus, and potash (NPK), 81
Norway, 90

Obama, Barack, 4, 25–26, 29, 90
Ontology, 16–20, 30, 49, 58, 116, 122, 125–131, 138–139, 145
Our World Is Not for Sale Network, 62

Patents, 50, 57, 59, 149
Paterson, Matthew, 107
Peasant/campesino, 5, 25, 30 40, 48, 50, 57–57, 72–73, 75, 79–81, 84, 88–89, 91, 94,118, 122, 132, 147, 149–150
Peltier, Leonard 136
Periphery. *See* World capitalist system
Philippines, 4, 93
Poulantzas, Nicos, 105
Polanyi, Karl, 95
Pollack, Paul, 151
Powder River War of 1864, 134
Prebisch, Raúl, 36
Precautionary Principle, 54
Pueblo peoples, 132–134

Rand, Ayn, 122–124
Red Cloud, 134. *See also* Lakota and Nakota
Reverse quarantine, 149
Rockefeller, David, 86
Rockefeller Foundation, 33, 81, 87–88, 90, 93, 140
Roman Nose (Cheyenne), 134. *See also* Lakota and Nakota
Rosebud, Battle of (1876), 135. *See also* Lakota and Nakota
Rousseau, Jean-Jacques, 11, 101
Russia, 2

Sacred thread (*Te Aho Tapu*), 130
Santa Cruz Declaration on Intellectual Property, 143

Schlosberg, David, 116
Scott, James C., 75, 96
Seattle, Chief, 57
Shadow banking, 124
Sitting Bull, 134. *See also* Lakota and Nakota
Sixth Great Extinction, viii, 98, 125–126
Slave-Sugar-Cotton triangle, 109
Social contract, 37, 94, 97, 101–102, 107, 115–116, 124, 150
South Africa, 89
Spain, 79, 109, 124
Spring Creek Declaration, 152–153
State of nature, 95, 97, 102
Stewart-Harawira, Makere, 121, 130. *See also* Maori
Structural Adjustment Participation Review Institution Network (SAPRIN), 41–44
Structural functionalism, 95
Suicide, 82
Survival International, 141

Thatcher, Margaret, 15, 21, 37, 46, 66, 122
Third World Network, 60, 141
Tinker, George, 137, 145–147
Tolstoy, Leo, 94
Traditional ecological knowledge (TEK), 87, 138
Transnational corporations, 5, 7, 9, 12, 14, 21, 24, 38, 44, 49–51, 62, 81, 114–115, 149
Tribalism, 6, 74, 95, 101, 127–129, 136, 141, 143
Turner, Ted, 86

United Kingdom, 28, 44, 92, 107
United Nations, 36, 58, 82, 143, 145, 152; Conference on Sustainable Development 2012 (Rio+20), 3, 4, 144; Declaration on Indigenous Rights, 91, 143; Development Program, 24; Food and Agriculture Organization (FAO), 48, 64; United Nations Conference on Environment and Development 1992, 121, 140; United Nations Conference on Trade and Development (UNCTAD), 36, 3–4; World Summit on Sustainable Development (WSSD), 143
United States of America: agriculture, 86, 89–90, 140; climate change and, 2; Congress, 1; greenhouse gas emissions (GHGs) and, 3, 5; indigenous peoples and, 101, 132–138, 141; international financial institutions and, 21, 23–29, 35–37, 44, 54–55, 59–61, 93; neoliberalism and, 35, 39, 45, 64, 68, 123; power and, 10, 70, 77, 92, 98; social movements, 119; torture, 9–10; war, 9

Urstaat, 96
US Agency for International Development (USAID), 90
US Department of Agriculture (USDA), 8, 84, 86, 90

Via Campesina, La, 73, 79–80, 89, 94, 122

Waken, 136–137. *See also* Lakota and Nakota
Wallerstein, Immanuel, 7, 14–16, 20, 65, 67–68, 70, 75, 81, 83, 99, 119
Washington Consensus, 45, 69, 73, 75, 79
Weber, Max, 101
Wildcat, Daniel, 65
Wilmer, Franke, 99
World Bank, 7, 11, 21, 23, 25–27, 31, 36, 40–41, 44, 48–49, 52–53, 60, 66, 80–81, 84, 87, 112, 151–152; Currie Commission, 33; Economic Development Institute (EDI), 33; International Centre for Settlement of Investment Disputes (ICSID), 53–54; International Development Association, 27. *See also* International Monetary Fund, structural adjustment programs
World capitalist system: core, 7–8, 12, 14–15, 18, 21, 29, 32–34, 36, 45, 47–48, 50–51, 53, 56, 58–64, 66, 69–70, 85, 88–89, 91, 96–98, 104, 109–113, 115–116, 119, 123, 126, 131, 146, 152–153; periphery, 7, 15, 29, 32, 35–36, 59, 62, 96, 99, 104, 109, 111–113, 115, 126, 140
World Economic Forum, 29, 70
World Indigenous movement, 74–75, 118, 135, 139–154
World Indigenous Peoples Conference on Territories, Rights and Sustainable Development. *See* Kari-Oca II
World Ocean, 1, 103, 108. *See also* Global warming
World peasant movement, 118
World Social Forum, 73, 75, 150
World Summit on Sustainable Development (WSSD), 143. *See also* United Nations
World Trade Organization (WTO), 11, 21, 23, 25, 49–63, 79–80; Aggregate Measurement of Support (AMS), 61; Agreement of Agriculture (AoA), 59–62; Agreement on Sanitary and Phytosanitary measures (SPS), 54; Codex, 54; General Agreement on Trade in Services (GATS), 52, 54–55, 59; Nonagricultural Market Access, 63; Overall Trade Distorting Domestic Support (OTDS), 61; Production and Processing Methods (PPMs), 54–55; Seattle protests (1999), 62, 79, 116; special safeguard mechanism (SSM), 63;

Trade-Related Aspects of Intellectual Property Rights (TRIPS), 57–59, 63. *See also* General Agreement on Tariffs and Trade (GATT), Uruguay RoundWorster, Donald, 107–108

Xane Indigenous group, 91

Yakama, 134
Yellow Bird (Lakota), 135

Zambia, 60
Zapatistas (EZLN), 78–79, 84, 97
Zimbabwe, 41–43, 87–88

About the Authors

Sharon J. Ridgeway received her PhD in Political Science from Northern Arizona University after spending twenty years in film production and theater in both Los Angeles and New York City. She is currently an Assistant Professor at the University of Louisiana at Lafayette, where she teaches courses on environmental policy and the role of mass media in democracy. She has settled in Lafayette, Louisiana, home of Festival International de Louisiane, great food, and great music.

Peter J. Jacques, like Sharon, holds a PhD in Political Science from Northern Arizona University, where he was first trained in environmental and indigenous politics. He is Associate Professor of Political Science at the University of Central Florida, where he teaches environmental politics, indigenous politics, and sustainability. He is currently the Managing Executive Editor at *The Journal for Environmental Studies and Sciences*. Peter has written many academic articles about environmental politics and sustainability as well as three prior books: *Environmental Skepticism: Ecology, Power, and Public Life* (2009), *Globalization and the World Ocean* (2006), and *Ocean Politics and Policy* (with Zachary A. Smith, 2003). He lives in Oviedo, Florida, with his beloved wife and daughters.